Collector's Guide

GRANITE PEGMATITES

Schiffer Earth Science Monographs Volume 8

4880 Lower Valley Road, Atglen, Pennsylvania 19310

Vandall T. King

Cover Photo: Tourmaline with Cleavelandite, Kurghal, Laghman Province, Afghanistan. FOV (Field of View) = 20 x 30 cm *Photo courtesy Stuart Wilensky*

Foreword from the Series Editor

This Schiffer Earth Science Monograph focuses on granite pegmatites, which are admired by gem and mineral collectors as a source of fine specimens, gem materials, and sometimes stupendously huge crystals. Pegmatites can also host important deposits of ore minerals and industrial commodities such as feldspar and mica. At the same time, their physical and chemical complexity of granite pegmatites present daunting challenges to researchers who seek to better understand their nature and origins.

Vandall King has devoted his career to the *practical* study of pegmatites in all their aspects: mineralogical, commercial, and historical. An experienced and internationally recognized author and expert, Van is uniquely qualified to make this complex topic interesting and understandable to advanced mineral collectors and laymen alike.

— Robert J. Lauf

Other Schiffer Books on Related Subjects:
Collector's Guide to Fluorite. Arvid Eric Pasto. ISBN: 9780764331930. $19.99
Collector's Guide to the Epidote Group. Robert J. Lauf. ISBN: 9780764330483. $19.99
Collector's Guide to the Mica Group. Robert J. Lauf. ISBN: 9780764330476. $19.99
Collector's Guide to the Pyroxene Group. Robert J. Lauf. ISBN: 9780764334047. $19.99
The Collector's Guide to the Three Phases of Titania: Rutile, Anatase, and Brookite. Robert J. Lauf. ISBN: 9780764332685. $19.99

Library of Congress Control Number: 2010931145

Designed by Mark David Bowyer
Type set in Bernhard Modern BT / New Baskerville BT

ISBN: 978-0-7643-3578-5
Printed in China

Schiffer Books are available at special discounts for bulk purchases for sales promotions or premiums. Special editions, including personalized covers, corporate imprints, and excerpts can be created in large quantities for special needs. For more information contact the publisher:

Published by Schiffer Publishing Ltd.
4880 Lower Valley Road
Atglen, PA 19310
Phone: (610) 593-1777; Fax: (610) 593-2002
E-mail: Info@schifferbooks.com

For the largest selection of fine reference books on this and related subjects, please visit our web site at **www.schifferbooks.com**
We are always looking for people to write books on new and related subjects. If you have an idea for a book please contact us at the above address.

This book may be purchased from the publisher.
Include $5.00 for shipping.
Please try your bookstore first.
You may write for a free catalog.

In Europe, Schiffer books are distributed by
Bushwood Books
6 Marksbury Ave.
Kew Gardens
Surrey TW9 4JF England
Phone: 44 (0) 20 8392 8585; Fax: 44 (0) 20 8392 9876
E-mail: info@bushwoodbooks.co.uk
Website: www.bushwoodbooks.co.uk

Contents

Dedication

The author is pleased to recognize two important people involved with gem-bearing pegmatite exploration, individuals who inspired a book that can be used by everyone who desires to understand gem-bearing granite pegmatites: Robert A. Brown and Louise Jonaitis. In their chosen professions, Bob and Louise have frequently met people who wanted to know more about granite pegmatites. Bob and Louise have generously shared their experiences with others so that an appreciation of gem deposits could reach as wide an audience as possible.

Louise Jonaitis is a founder and board member of Plumbago Timber and Quarries LLC, with its principal gem pegmatites on Halls Ridge, Newry, Maine. Louise has had a lifelong involvement in various aspects of the mineral and gem industry and has been a wholesale marketer and distributor of gemstones. She has worked closely with gem cutters regarding quality, utility, and value of gem rough, particularly gem tourmaline. As an active land agent, she has been involved in acquisition and utilization of various properties for residential, logging, mining, and other purposes. Louise has been concerned in the long-range plans in many land development activities. Louise has experience drilling

Louise Jonaitis

dynamite holes as well as the operation of heavy mining machinery and has been mentored in mining practices and exploration techniques.

Robert Brown, of Hanover, Maine, is a fifth generation miner. The particular ridge where the graphite occurred is called Scriba Mountain, indicating that scribes would use graphite. In 1972, Bob was in charge of security at the Dunton Gem Quarry, a property he now owns. Bob has a lifelong experience in land-related professions and activities, not just mining, but also in excavating, logging, etc. Bob is proficient in drilling techniques and has extensive experience with mining equipment. In 2004, he formed Plumbago Timber and Quarries LLC in fulfillment of the long-term plan to develop the Halls Ridge gem pegmatites. As part of this goal, Bob extended the contiguous Brown Family land holdings from U.S. Route 2 in Hanover to State Route 5 in Rumford. These lands encompass all of the well-known gem-bearing pegmatites on Halls Ridge and parts of Plumbago and Puzzle Mountains.

Robert Brown

Acknowledgments

The author is very grateful to many people who have helped make this book a reality. Miners have allowed their quarries and work to be photographed. Mineral collector and dealers have permitted photographs to be taken of their specimens. Friends and colleagues have openly shared their experiences and knowledge with the author. In particular, many people have supplied images that greatly improve the illustration quality of this book. Please note additional acknowledgements in the photograph captions.

Warm and hearty thanks are due to: John Barlow, Russ Behnke, Dudley Blauwet, Robert Brown, Jan Brownstein, Priscilla Chavarie, Roger Clapp, Jeff Collins, Katie Collins, Val Collins, Larry Conklin, Anne Cook, Bill Cook, Rock Currier, Bob Daigle, Kevin Downey, Dennis Durgin, George Elling, Jeff Fast, Jesse Fisher, Wayne Flanders, Gene Foord, Gary Freeman, Mary Freeman, Joseph Freilich, Zeus Freilich, Robyn Greene, Elna Hauck, Richard Hauck, Barry Heath, Leonard Himes, Dennis Holden, Ron Holden, Gary Howard, Louise Jonaitis, Paul Kelley, Nathaniel E. King, Brian Kosnar, Ron Larrivee, Bill Larson, Will Larson, Rob Lavinsky, Bryan Lees, Dona Leicht, Wayne Leicht, Luiz Alberto Menezes (Filho), Janet Nemetz, Herb Obodda, Monika Obodda, Jane Perham, Frank Perham, Dan Schwind, Gail Spann, Jim Spann, Tim Swan, Wayne Thompson, Woodrow Thompson, Dan Weinrich, Stuart Wilensky, Stephan Wolfsried, Ray Woodman, and James Zigras.

Preface

Rubellite crystals on albite. 3 x 4 cm. Malkhan District, Russia. *Jeff Fast Collection.*

This book is written for the person wanting to know about granite pegmatites in uncomplicated terms and is not intended for the specialist wanting to know all of the latest scientific models or theories. This book will contain examples from North America whenever possible, although worldwide crystal specimens will illustrate the text. The content of the book will be general and particular kinds of pegmatites might receive almost no mention, but the general ideas about them will be the same.

Geological theories developed very differently because an eighteenth century German geologist, Abraham Werner [September 25, 1749 – June 30, 1817], lived near some very confusing rocks and he, nonetheless, developed a theory for their origin. Werner was a popular teacher and his students went out into the world believing that all rocks were formed from an ancient ocean that covered the world at one time. When his students tried to study rocks all over Europe, they discovered "fire," that is volcanic action. The new Volcanic Theory was too compelling and Werner's ideas soon became historical curiosities. Few of Werner's students held onto his teachings after the controversy concerning the origin of the primeval rocks was "settled," although geological research papers were lively reading on the subject from the 1790s through the 1820s. Both insults and "fur" flew in print regarding the origins of rocks. Certainly there is still enough lively reading on the origins of rocks, today, although the current texts have a more gentlemanly air. The difference in today's research articles is a matter of detail. There is much agreement over the "big picture," but the study of rocks and minerals is alive and well.

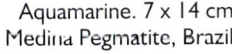

Aquamarine. 7 x 14 cm. Medina Pegmatite, Brazil.

One of the world's most interesting rock types is called granite pegmatite. It has some minerals like those in granite and has giant crystals along with tiny crystals that were formed at about the same time. The word "pegmatite" indicates the close association of large and minute crystals. The forces that were at work to make pegmatite texture are very uncommon in earth's history and the research into why pegmatite texture exists is a life-long study. This book will contain a minimum of hard science because it seeks to fill a void in the literature. The distinction may be that this book discusses "what is" rather than "what if." The research into granite pegmatites is now a field dominated by laboratory studies wherein there is the hope that experiments directed at synthesizing rocks will support a theory of what has been observed in nature. The intended audience of this book seeks to know about granite pegmatite and nature's finest and most beautiful crystals.

If the reader is interested in knowing what is currently believed about the origins of granite pegmatites, this author can recommend no better book than David London's *Pegmatites*. London's book is full of reviews of recent scientific research and discusses how that research supports or implies that certain conditions observed in the field have been duplicated in the laboratory and therefore the process of granite pegmatite formation may be understood. His book is aimed at the professional geologist and the scientifically trained enthusiast. Mercifully, we still live in interesting times. London freely admits there are many questions that are waiting to be answered and many existing experimental results require further substantiation. Books such as *Internal Structure of Granitic Pegmatites* (Cameron et al., 1949) will remain fresh and relevant because the subject is treated from the perspective of observations. It is true that there are new observations being made, but for the most part, the observations are restricted to minutiae. The chaos about origins that characterized Werner's research and theories is largely gone in modern studies. Nonetheless, anyone interested in knowing the scientific details concerning the origins of granite pegmatite and the many intricacies of scientific research will have to have a good background in chemistry, physics, and mathematics, as even "introductory" level books are written with the college student in mind.

The point of view of this book will be practical. There are discussions about mining, minerals, and observations that can be made to better understand the successful path in searching for minerals in a granite pegmatite. The mineral collector and the miner need to "read" the rocks to find the hidden treasures. Mineral naturalists maintain reference and display collections and, most importantly, it is still possible to explore and discover your own materials. Some naturalists, such as ornithologists, find it impractical if not unwise to have a collection of birds and they prefer note taking, photography, and writing as an outlet for their interests. Mineralogists and paleontologists realize that, unlike the living world, the world of rocks and minerals is non-renewable. Erosion is a constant threat to minerals and rocks that are on or near the surface of the Earth. Mineral conservation is the realm of the mineral collector and that which is not preserved will surely be destroyed by erosion without replacement. This book is written for naturalists and collectors.

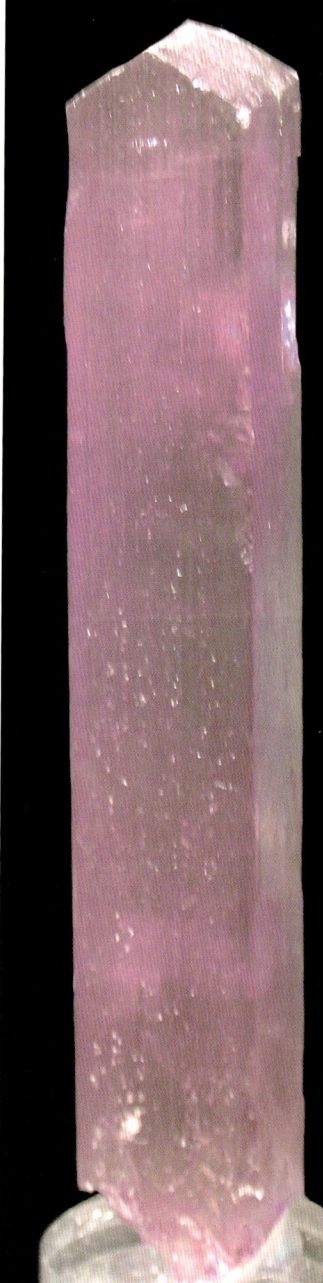

Kunzite. 14 x 3 cm.
Mawi, Afghanistan.

Chapter One

Granite, Granite Pegmatites, and the Search for Gems

Granite pegmatites are very special rocks and this book is designed to introduce them from the point of view of observation. The technicalities of "Why" will be left to scientific articles. From a miner's or mineral collector's point of view, descriptions are the first step to discovering great minerals in these rocks. Expectations of what minerals may be found in granite pegmatites abound and it is important to remember that only about 25% of the known minerals have been found in these rocks. Gold has been reported from a few granite pegmatites, but only as "curiosity amounts" best viewed by a microscope. Specimens such as a crystallized gold nugget have yet to be found in them.

Gems are well-known in granite pegmatites, but when one reads of opal from a granite pegmatite, it must be remembered that precious opal, such as found in the famous opal fields, will not be found. Of the species found in granite pegmatites, few produce gemstone-producing crystals. The list of the frequently desired gem-producing crystals includes: almandine, apatite, beryl (including aquamarine, golden beryl, goshenite, heliodor, morganite), chrysoberyl, elbaite, liddicoatite, montebrasite, oligoclase (peristerite), petalite, quartz, rossmanite, spessartine, spodumene, and topaz. After these few species, the rarity of additional gem minerals increases dramatically. Some minerals, such as lepidolite, are primarily mined from granite pegmatites and may be used for ornaments. Nonetheless, miners are always on the alert to recognize rare gems if they present themselves. One of the most common minerals in granite pegmatites includes the feldspar group. Uncommon look-alike ornamental materials in the group are peristerite and moonstone. Some pegmatites produce great quantities of these minerals, but most districts yield none at all.

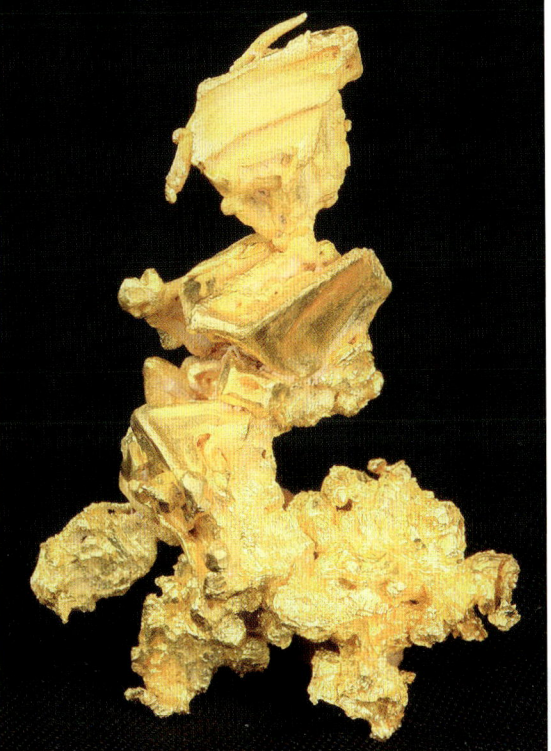

"The Scholar" gold nugget. 32 x 21 x 17 cm. Colorado Quartz Mine, Mariposa Co., California. *Photo courtesy Joseph Freilich.*

Certainly, tourmalines are the most important gem crystals from granite pegmatites. The important fact to remember is that gem crystals usually look like gems. Over-enthusiastic miners have dreamed their highly fractured "gem" crystals would yield them a fortune in gemstones, until they discovered that every flaw or fracture represented waste. "Gem" crystals, which really are too flawed to yield true gems, may still be prized for museum-style displays are another matter and individual specimens may yield several years' good wages in the marketplace.

Elbaite (Tourmaline Group). 7 cm tall. Paprok, Afghanistan.

Opal (Precious). 2 cm wide. Coober Pedy, Australia.

Very successful mining ventures have occurred when a miner or company have revisited a previously exposed deposit under the theory that there is always something left behind. If the previous miners did not "read the rocks" carefully, treasures may indeed remain. However, miners with a good background in understanding the deposits they worked are less likely to have left much behind of value. Sometimes, there are considerable high quality minerals remaining, but the miners recognized that the cost of mining could exceed the value of the remaining minerals. For these and other reasons, people who are serious about discovering minerals need a background in the deposits of interest. Past history is useful, but the future is the miner's concern.

Moonstone. 2 x 3 cm. Rabb Canyon Pegmatite, Grant Co., New Mexico.

Peristerite gem. 1.5 x 2 cm. India.

Staying with the example of tourmaline, there are many species and varieties in this group. A variety differs in some way from others of its species by virtue of insubstantial aspects such as a different color produced by a trace component or unusual growth patterns, etc. The colors of tourmaline have been given a variety names that otherwise sound like species names. Red and pink tourmalines are called rubellite, although the original use of the name was only for the beautiful deep red shades. Green tourmaline is called verdelite, while blue tourmaline is called indicolite. Achroite is colorless tourmaline. There are also variety names that are merely descriptive: black,

cinnamon, blue cap, etc. Watermelon is one of the most sought after tourmaline varieties. Watermelon has concentric color zones in the crystal: a green "rind" and a pink or red "core." Although this kind of tourmaline was known for a hundred years before, principally in North America, it was not named until 1910 when a naturalist, George R. Howe, named watermelon tourmaline based on crystals from the Havey Pegmatite in Poland, Maine, USA. Howe also introduced the variety name "cucumber" tourmaline for green crystals with a white core, but that combination is rare in good examples. Variety names based on color may actually be given to several different species.

Mixed lot of gem tourmaline: green and red tourmaline is from Dunton Quarry, Newry, Maine, while the small red crystals are from Mount Mica Quarry, Paris, Maine. FOV (Field of View) = 9 x 4 cm.

Achroite and pale rubellite zones in Elbaite. 4.4 x 1 cm. Mount Mica Quarry, Maine.

Blue cap elbaite, Tourmaline Queen Pegmatite, Pala, California. 12 x 7 cm.

Black Tourmaline (Schorl). 11 x 7 cm. Shengus, Pakistan.

Schorl. 7 x 5 cm. Barnet Locality, Canton, Maine. *Norman Davis Collection*.

Watermelon tourmaline (elbaite). 5 x 6 cm. Dunton Pegmatite, Maine.

In granite pegmatites, green and blue tourmalines may be elbaite or even schorl while red and colorless tourmalines may be elbaite, rossmanite, or olenite. Elbaite is the most common species of the brightly colored tourmalines in granite pegmatites, but worldwide, the number of colored tourmaline includes many more species. Black tourmaline in granite pegmatites is usually schorl, but research is showing that another black tourmaline, foitite, is not rare.

Beryl also has a large number of varieties. The oldest name, of course, is emerald. Emerald has been known since ancient times and the name has been given to a variety of similarly appearing minerals and, of course, has had many names depending on local languages. Beryl was also an ancient name and was used for less beautiful material, but the fact that emerald and beryl are the same basic mineral was not known until about 1790. The name emerald was so well established as a mineral name that it continued to be used as the preferred species name by some authors even until the mid-1800s. Aquamarine, signified the color of the sea, but it was first applied to the green colored mineral when it was named in 1747. As more blue gemmy beryl crystals were discovered, the name aquamarine was extended to include two colors. Yellow beryls were simply called golden beryl. There are two modern beryl varieties. Morganite is pink and heliodor is ideally golden orange, but heliodor is increasingly used as a synonym for golden beryl.

Rubellite with green termination. 6.5 x 11 cm. Paprok, Afghanistan.

Verdelite. 12 x 5 cm. Mount Mica, Paris, Maine, USA.

Indicolite. 2 x 4 cm. Darra-i-Pech, Afghanistan.

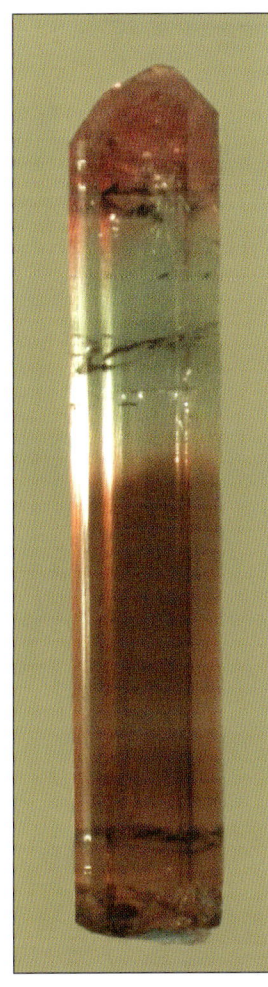

Bi-color tourmaline with central achroite zone, 1 x 3.5 cm. Coronel Murta Pegmatite, Brazil.

Watermelon tourmaline. 17 x 7 cm. Paprok. *Andreas Weerth Collection*.

Cinnamon tourmaline. 5 x 8 cm. Xanda Pegmatite, Virgem de Lapa, Brazil.

Pale pink beryl was discovered in Elba, Italy, in the early nineteenth century, but the crystals were small. A short time later, in the 1840s, pale pink non-gem beryl was found in granite pegmatite boulders in Massachusetts and named goshenite for the nearby town, but most of the beryl from Goshen was white to colorless and the name gradually became associated with the colorless mineral because collectors wanted to have goshenite in their collections and so disregarded the original pink requirement. "Cesium beryl" was named in 1884 for a white to colorless beryl found at Tubbs Ledge Pegmatite in Norway, Maine, USA. Many pink gem beryls have later been found to contain at least a little cesium and, because of that fact, the name "cesium beryl" has persisted when referring to any lightly colored gem beryls with good crystal form from gem pockets in granite pegmatites. In 1911, spectacular pink gem beryl crystals were found in San Diego County, California, and morganite was named in honor of James Pierpont Morgan, financier as well as a gem and mineral collector and benefactor of the New York's American Museum of Natural History. Interestingly, morganite frequently starts out its discovery life with a peach to orange coloration, but the orange tints may be bleached by sunlight and pink remains as a relatively stable color afterwards. In 1912, one of the most recent variety names of beryl was coined: heliodor. Heliodor was found in Rössing, Namibia, and a name given to extraordinary orange beryl that was being found by a German gem mining company. Also in this time period when beryl variety names were being coined, a bright red beryl from Topaz Mountain, Utah, was named bixbite for Maynard Bixby, a mineral dealer from Utah. There are separate species related to beryl. One is called bazzite and which rarely forms light blue crystals larger than a few millimeters and another species called pezzottaite found in Madagascar in 2002 and named for mineralogist Federico Pezzotta. This mineral is not a variety and has a bright raspberry red color. Cesium is not a coloring agent of morganite or pezzottaite.

Pezzottaite, Sakavalana Pegmatite, Madagascar. 6 x 1.3 x 0.6 cm.
Photo courtesy Rob Lavinsky.

Goshenite. 4 x 2 cm. Shengus, Pakistan. *Photo courtesy Joe Freilich.*

Aquamarine. 10 x 4 cm. Dassu, Pakistan. *Photo courtesy Joe Freilich.*

Aquamarine. 12 x 5 cm. Barra de Salinas, Brazil.

Orange Morganite. 15 x 15 cm. White Queen pegmatite, Pala, California, USA.

Heliodor. 2 x 5 cm. Serra de Mesa Pegmatite, Brazil.
Photo courtesy Joe Freilich.

Upper Right: Golden Beryl. 3 x 7 cm. Volodarsk-Volynskii, Ukraine.
Photo courtesy Joe Freilich.

Lower Right: Bixbite, 3.5 x 3 x 3.4 cm, Violet Claims, Utah. *Jim and Gail Spann Collection. Photo courtesy Jeff Scovil.*

Emerald. 1 x 2 cm. Muzo, Colombia.
Photo courtesy Jeff Collins.

How are Granite and Granite Pegmatites Related?

The definition of a rock is based solely on the kinds of minerals and proportions of minerals that are present. Granite must contain at least 10% quartz and at least one kind of feldspar making up for the remainder of the rock. There are also very simple granite pegmatites, which contain only these two minerals. True granites are generally composed of 40-80% feldspar and 20-60% quartz and other minerals may be present in a small amount, but the additional minerals are not essential to the definition of granite. A complication is that there are several kinds of granite. "One feldspar" granites are less common than "two feldspar" granites. The one-feldspar granites contain potassium-rich feldspar, such as microcline. Two-feldspar granites contain potassium-rich feldspar and plagioclase feldspar, but when plagioclase is the dominant feldspar, the rock is no longer a granite by definition. Plagioclase-rich rock, with at least 10% quartz, is called granodiorite. Even more technically, a one-feldspar granite is called an alkali-feldspar granite, and the two-feldspar granites are the real granites. The currently accepted nomenclature of rocks is governed by International Union of Geological Sciences (IUGS). The IUGS rules of naming rocks is sometimes called the Albert Strekheisen Classification named for the man responsible for the committee that established the new definitions, although the new classification is not all that different from older classifications. There are also variations according to mineral content: one-mica granite, two-mica granite, hornblende granite, etc., but they are all "granite."

Pegmatite is not a specific kind of rock, but is a kind of texture of a rock. Granites may be fine-grained, medium-grained, coarse-grained, porphyritic, orbicular, rapakivi, aplitic or simply aplite, graphic granite, and, of course, pegmatitic granite.

Fine-grained granites have very narrow size limits of their crystals ranging from fractions of a millimeter to a few millimeters. Fine-grained granites are prized as rocks for monuments and may have a salt and pepper look if they also contain dark minerals. The quartz and feldspar in coarse-grained granite may be several centimeters across, but there are few, if any small grains, for example measuring several millimeters. Interestingly, curbstones often have pegmatitic veins (to about 10 cm) in them.

Porphyry is also a word describing grain size texture. Granite porphyry may have large crystals, typically feldspars to several centimeters, in contact with small grains of feldspar, but there are no intermediate size grains. Rapakivi is an uncommon texture seen in coarse-grained or porphyritic granites. The feldspar grains are concentrically zoned, usually with a white plagioclase rim with a pink microcline feldspar core. Additional rare kinds of granite exist.

Fine-grained Granite (2x2 cm) with quartz (gray), feldspar (white), and biotite (black) from a curbstone.

Simple granite pegmatite in contact with typical fine-grained granite from a curbstone. 15 cm thick. The grains in the granite are about 1-2 mm, while the grains in pegmatite are sometimes over 1 cm.

Much of the geological and mineralogical literature simply uses "pegmatite" without specifying the kind, but almost always the most common full term granite pegmatite is implied. However, there are gabbro pegmatites, syenite pegmatites, nepheline syenite pegmatites, and more, just as there are gabbro porphyries, syenite porphyries, nepheline syenite porphyries, etc.

Pegmatites have grain sizes larger than normally encountered in most rocks. Grain sizes in most igneous rocks only get to several millimeters, occasionally centimeters, but in granite pegmatites grains to several tens of centimeters are almost "normal" and many crystal grains reach proverbial giant sizes. In fact, giant crystals to over 30 meters long have made pegmatites famous, but the word pegmatite actually means that there may be large crystals of a mineral next to tiny, small, and/or medium-sized crystals, all of the same kind. Most rocks form with all of their crystals within a narrow size range. Pegmatites are the opposite. Pegmatites

crystallized from a fluid and because of the unusual nature of the fluid, crystals of the same mineral grew at varying rates and times. The variable rate of pegmatitic texture is different than the two stages implied by a porphyry, where early crystals grew and a second generation filled in around the early crystals. Coarse texture implies a single growth event where relatively few crystals were able to grow more or less uniformly large. An important fact to remember is that crystal growth is a change from equilibrium. By definition, at equilibrium, growth is forbidden, because there is no change. In contrast, pegmatites show frequent change, many times in stages. Proportions of minerals may change; different kinds of minerals may begin to crystallize, or some minerals no longer form during later stages of growth. The seeming wildness of granite pegmatites is their excitement. While there are recognizable patterns in the growth of minerals in a pegmatite, there is also unpredictability.

Granite porphyry. Porphyries typically have large crystals surrounded by a groundmass of smaller crystals, which are more or less the same size. The analogy is a group with many short people and many tall people, but no one present with intermediate heights.

Coarse-grained granite. 12 x 15 cm. Rockville, Minnesota.

Pegmatite

Mineralogist and geologist René Just Haüy coined the word "pegmatite" in the early 1800s and his protégé, Alexandre Brongniart, publicized the word in 1813. The Greek root, *"pegma"* simply means, "something fastened together." Brongniart also used another term, *"granit graphique,"* for an intergrowth of quartz and feldspar commonly found in the granite pegmatites of central France. For many years there was an inconsistent use of the two terms, frequently used as synonyms. The Limoges granite pegmatite district in central France, which supplied material for the famous porcelain industries of that country, had much graphic granite. Jahns (1955) indicated that pegmatite became the word of choice in Europe by the 1850s. However, the name pegmatite didn't become widely used in the USA until about 1890 with previous USA geological reports only signifying the coarseness of granite veins or formations that also contained a wide variety of unusual minerals. Nonetheless, graphic granite has retained its original place in describing the specific quartz and feldspar intergrowth.

The variability of pegmatite texture is difficult to picture and there will be numerous examples of this texture in the illustrations of this book. When there are large grains and every size in between down to tiny grains, the texture has a somewhat chaotic appearance and this impression is somewhat true. While granite is often prized for monument work as every piece of a particular granite should be like every other piece, there are few if any two pieces of pegmatite that resemble each other.

Another problem with adequately describing granite pegmatites is related to their huge grain size. The examples of granite textures, except for pegmatitic texture, show that a representative piece of the rock can be enclosed in a small volume, perhaps a block only several centimeters on an edge, depending on the texture. Granite pegmatites are so variable in grain size and distribution of minerals that there is hardly a portion that could be chosen to represent the mineral percentages of the pegmatite. A few petrologists have attempted to find truly representative samples of pegmatites so that a chemical analysis could be made of a pegmatite. The most common analyses have about 71-76 weight percent silica (SiO_2) and 13-16 weight percent alumina (Al_2O_3). Soda (Na_2O) may be 2-5%, while potash (K_2O) may be about 4-8%. Lime (CaO), iron oxides (FeO and Fe_2O_3), and other constituents are usually less then 1% each. However, pegmatites are known also for concentrating rare elements and having rare-element-bearing crystals sometimes to tens of kilograms. However, the rare-elements may have constituted only a small percentage of the entire original pegmatite-forming fluid.

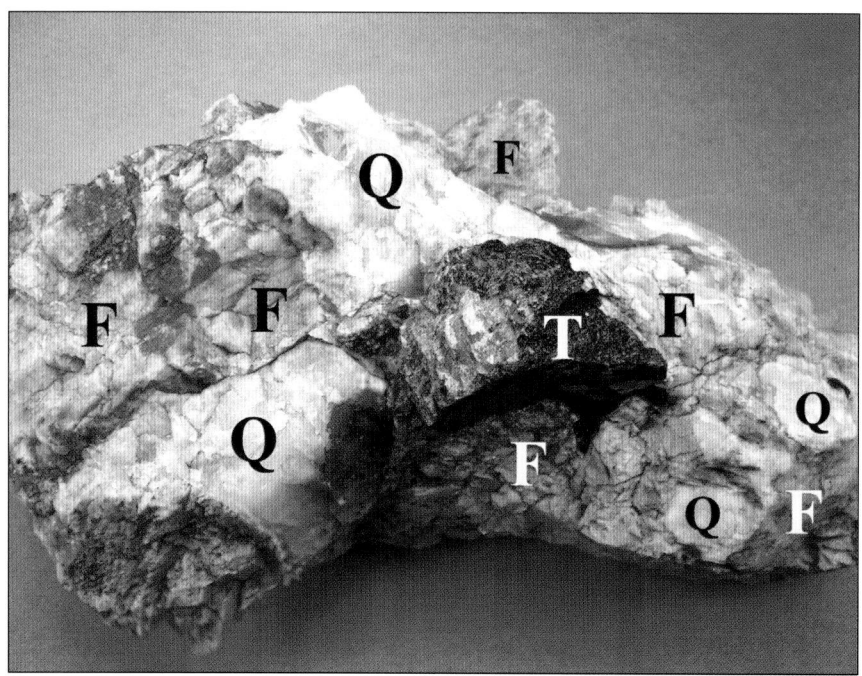

Typical pegmatite. Q = quartz, F = feldspar, T = black tourmaline. 8 x 10 cm. Bumpus Pegmatite, Maine.

Aplite

Aplite is a rock texture commonly associated with granite pegmatites. Aplitic texture is very uniform and is believed to represent sudden crystallization distributed over many crystals. Aplite texture has been called "sugary" because the rock looks as though it is composed of granulated sugar grains cemented together. In some places, such as in the granite pegmatite districts in San Diego County, California, aplite may form repetitive layers called "line rock" and may be used as an indicator of favorable crystallization in a pegmatite. Aplite may form sheets separating pegmatite from a crystal pocket or replacement unit or may form on both sides of crystal pockets. Aplites may be present without being next to replacement bodies or crystal pockets. There is some argument about how rapidly granite pegmatites may form, but aplites certainly formed extremely rapidly in comparison.

Black tourmaline trapped in rapidly formed Aplite. 12 x 12 cm. Bennett Pegmatite, Maine.

Aplite with lens of blue tourmaline-quartz pegmatite. 15 x 18 cm. Crooker Gem Pegmatite, Maine.

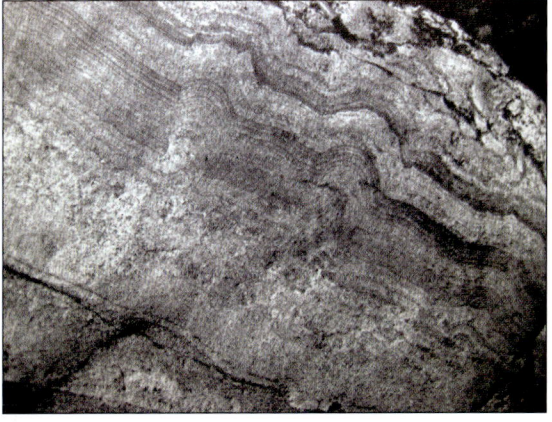

Layer Aplite line rock with dark garnet layers, FOV = ~1 x 1 m. Pala, California. Jahns (1955).

Aplite in bottom right cutting granite pegmatite, FOV about 05 x 0.75 meters. Howard-Collins Pegmatite, Maine.

Aplite with blue tourmaline, quartz, and cleavelandite lens. 10 x 14 cm. Crooker Gem Pegmatite, Maine.

Graphic Granite

Graphic Granite may be considered to be a special kind of granite pegmatite. The name graphic granite has survived as a textural term because this intergrowth of quartz and microcline feldspar is so common. The pattern of graphic granite seemed to mimic ancient writing and this texture has been given many names including "runite" because the pattern looked like runes or runic writing. The pattern is considerably variable and an important feature is that quartz forms rods, which may persist for many centimeters into the host microcline. Adjacent masses of graphic granite may have very different patterns and grain sizes. The blocks of feldspar may consist of either microcline or albite, although graphic granite with microcline is by far more common. Masses of graphic granite may be a number of meters in all dimensions. Graphic granite is not always present and some pegmatite districts contain it in great minable quantities, while it may seem rare in other districts. Some pegmatite districts mined graphic granite exclusively, although it contained considerable quartz and was worth less than high-grade feldspar. The porcelain industry was very particular in buying ground feldspar with as little quartz as possible, but ground graphic granite could be used as mineral filler, grit, etc. Graphic granite occurs relatively early in the crystallization sequence of granite pegmatites and is considered a poor sign for the kind of granite pegmatite which yields gem pockets, but many crystal specimens are found with graphic granite and crystal pockets may be found where replacement has cut graphic granite. Usually, graphic granite means there is a lot of rock that needs to be moved before valuable minerals could be found. Most granite pegmatites are barren or nearly barren of crystal pockets.

Graphic granite with with well-developed geometrical pattern, 5 x 10 cm, Tryon Mountain Pegmatite, Pownal, Maine.

Graphic granite. Poorly geometrical pattern. 8 x 10 cm. Bumpus Pegmatite, Maine.

Graphic granite. Equidimensional pattern. 8 x 8 cm. Keith Pegmatite, Maine.

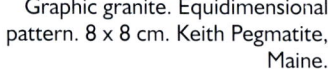

Giant Crystals

The definition of pegmatitic texture reveals nothing about the size limit of big crystals that may be found in a pegmatite. Spectacular granite pegmatites may have masses of microcline over 30 meters in maximum dimension along with microcline grains a few centimeters to tens of centimeters, including microcline grains down to a few millimeters, all roughly next to each other.

Some minerals commonly form big crystals in pegmatite. Beryl crystals up to 1 meter are so frequent as to rarely make it to the record books. Most such crystals grew in competition with other minerals, but were able to grow fast enough so that their crystal shape was expressed. When other minerals crystallize and begin to encase the growing beryl, the shape of the crystal may be affected. The faster growing beryl has a taper with the first formed part of the crystal being the smaller end. The termination part of the crystal expanded into the fluid, which was crystallizing. Beryl crystals found in 1928 at the Bumpus pegmatite, Albany, Maine, were up to 18 feet long and 4 feet (5.49 x 1.22 meters) in diameter. Eventually, beryl crystals were found at the Bumpus Pegmatite that were up to 33 feet long (10.06 meters). A giant beryl reported from Malikialina District, Madagascar was supposedly an astonishing 18 x 3.5 x 3.5 meters, but despite its reputed size no one photographed it and no one wrote a contemporary report about it. It was a personal communication to a person looking for giant crystal records (Rickwood, 1981), but because of the absence of a primary description or a specific locality, the Madagascar record may be apocryphal (Claus Hedegaard, personal communication, 2008).

Quartz crystals also have exceptional size in pegmatites, but quartz crystals form in gem pockets or crystal cavities. Pegmatites in New England and in southern California produce quartz crystals that are huge or which have been prized for their great beauty and many crystals weighing over 100 kg have been found, but amazing crystals to nearly 20 meters have been written about from worldwide areas, but no photographs are known. The largest photographed quartz crystals seen by the author are presented in this section.

Gem green tourmaline crystals with quartz crystals on cleavelandite, about 1 x 1 meter, Pederneira Pegmatite, Brazil.

Giant beryl crystals with Wallace Cummings and unidentified miner. FOV = 5 x 5.5 meters. Bumpus Pegmatite, Maine. *Photo courtesy Benjamin Shaub.*

The most famous photographs of giant crystals show crisscrossed spodumene crystals in a zone. The Harding Pegmatite in Taos County, New Mexico, and the Etta Pegmatite in Pennington County, South Dakota, have perhaps the best examples of giant spodumene.

What Sizes and Shapes Do Granite Pegmatites Have?

Granite pegmatite bodies may be of many sizes. Virtually every street paved with granite curbing shows pegmatites cutting across a few of the curbstones. Granite pegmatites have frequently been called veins. They are also called dikes because they resemble "dike walls" dividing the main mass of the rock. Granite pegmatites may be very vein-like in surface exposure, but they are sometimes not very much longer than they are wide and then they are said to have a "lens" shape, because they frequently look like a concave lens in cross-section. Most of the pictured examples are relatively small by granite pegmatite standards. Some gem-bearing pegmatites in San Diego County, California, or in the Haramosh Mountains area of Pakistan are only one to two meters thick and have been traced for up to a hundred or more meters in maximum length and up to about the same distance in width. Because the length and width may be about the same, when pegmatites are discussed with a third dimension, the "dikes" are visualized as "sheets" cutting the larger mass of rock.

Van King standing between giant quartz crystals from Madagascar.

Harding Pegmatite, New Mexico showing central zone of crisscrossing white spodumene bladed crystals 3-4 meters long. Width of view is about 40 meters. From Jahns (1953).

Many economic mica-bearing or feldspar-bearing pegmatites are up to 100 meters thick and several hundred meters long although some big pegmatites extend over a kilometer. In both California and in Pakistan and other areas, relatively thin pegmatites have yielded great amounts of outstanding gem materials. The percentage of "fertile" pegmatites is low in a region. In some instances, there are no pegmatites of economic or gem interest in an area. Even in famous pegmatite fields with gem-bearing pegmatites of international acclaim, there may be only a few pegmatites worthy of notice among the numerous deposits that have yielded commercial quantities of feldspar, mica, or beryl. Oftentimes, more value has been acquired from mining and selling tons of industrial grade minerals than from the search for more elusive gems. In some areas, gravel pits have been the most valuable deposits in the region, if there is a ready market nearby such as a prosperous town or city. Gem fever is as real as gold fever and the best a miner can do is choose as wisely as possible.

Detail of giant spodumene zone. View about 1 x 1 meter. From Jahns (1953).

Pegmatites in Three Dimensions

All of this vocabulary is more or less informal. One report will discuss pegmatite dikes, while another may only refer to pegmatite sheets. In most situations, the actual shape of a pegmatite is not apparent, except in areas where there is well-exposed bedrock without soil cover or vegetation. The following diagrams show that whatever is exposed in the surface is sometimes unrevealing of what changes may be deep inside the rock. When a granite pegmatite is exposed in an outcrop, a miner's decision to dig is a difficult one. The accompanying illustration of a pegmatite shows some black tourmaline, but the exposed pegmatite is only centimeters thick. The curbstone exposure shows how a thin pegmatite may look very thick because the dike intersects the surface at a shallow angle. The surface exposure may be very deceptive. One example looks small, but appealing, while another has a large surface area, although it really is a thin dike. Any splits and underground channels are also hidden. An abrupt change in the angle of the dike is one feature that is always looked for in pocket-bearing granite pegmatites. A horizontal contact in an otherwise dipping vein might be a place where pocket-forming fluids were trapped in their upward migration and the change in structure may be where the best minerals formed. When a granite pegmatite is being prospected, it is one thing to be enthusiastic and quite another to be sure. The limits of the pegmatite need to be traced, although many pegmatites have little contact rock visible at the surface and much may be hidden by vegetation and dirt. The pegmatite surface area needs to be as clean as possible to assess the profitability of mining.

The ledge shows a tapered light colored granite pegmatite dike or vein on the bottom. The central portion of the ledge shows light colored granite pegmatites with different amounts of regularity. The upper left dike resembles a convex lens in cross-section while the two pegmatites in the center show irregular pinching or tapering. Field of view is about 2 x 3 meters. New Gloucester, Maine.

Granite pegmatite in granite exposed in a large road cut. The pegmatite has aspects of a vein in cross-section, but has a bulge that might lead some to call it a lens, but in reality the pegmatite extends deep into the rock and is probably sheet-like. Note black tourmaline. View about 1 x 1 meter. New Gloucester, Maine.

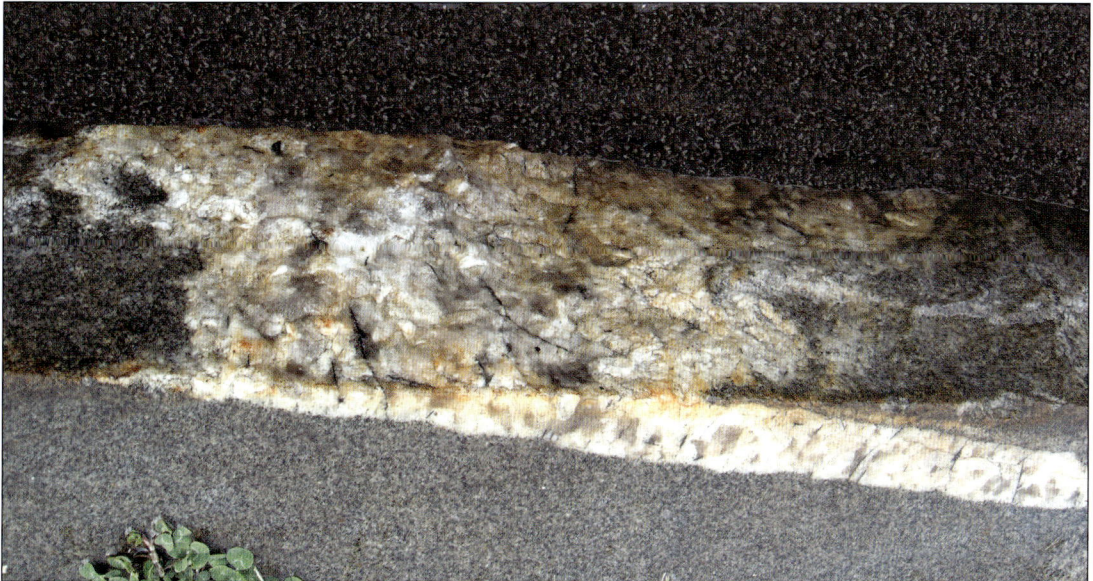

Shallow dip angle of a pegmatite may give a false impression of true thickness of the pegmatite. From a curb stone. 15 cm thick. Looking at the top of the curb (lower center), rough roadside face of curb (center), and road surface (top).

Granite pegmatite exposed in a cross-section of the "Earth." FOV = 8 x 12 cm. *Photo courtesy Nathan King*.

Intersecting pegmatites at the "surface" are actually a single channel at depth. FOV = 8 x 12 cm. *Photo courtesy Nathan King*.

Pegmatite showing a symmetrical kink or offset in a dike. *Photo courtesy Nathan King*.

Pegmatite dike showing a split below the "surface". 8 x 12 cm. *Photo courtesy Nathan King*.

Distribution of Granite Pegmatites and Their Exploitation

Worldwide, pegmatite districts are abundant. Some countries, such as Brazil and Madagascar, seem to have gem bearing granite pegmatites wherever one might search. Some countries, such as Italy, have a few choice pegmatites. On the scale of a world map, districts do not particularly show up without exaggeration of symbols and it is not essential to show every single concentration of granite pegmatites. Also, symbols on a map do not always convey quantities of gems produced or when their time of peak production occurred. The gem-bearing pegmatites in the Ural Mountains were active in the eighteenth century while the Brazilian gem pegmatites just began to be well known by the late nineteenth century, but gem production dramatically rose in the 1950s.

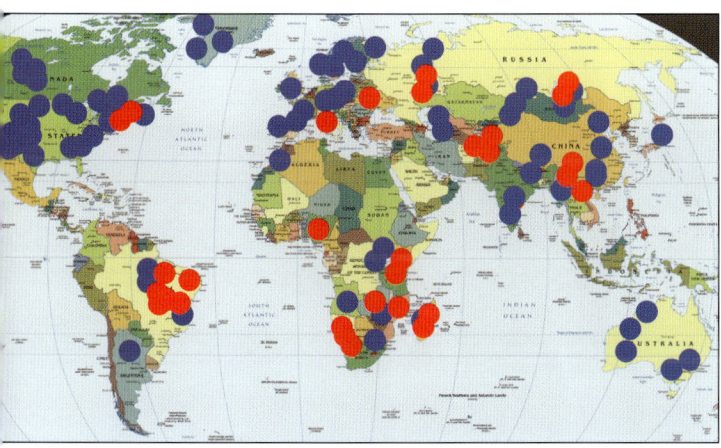

Pegmatite Fields or Districts worldwide. Red dots indicate areas particularly known for gem pockets while blue dots indicate important areas for pegmatites in general.

Pakistan and Afghanistan began to be famous for their gem pegmatite minerals in the later part of the twentieth century and remain strong producers. In North America, Maine was famous for a relatively small production of gems in the later nineteenth century, but California became prolific in the early and late twentieth century with respect to granite pegmatite gem crystals. Brazil has undoubtedly produced the largest quantity of gem crystals to date, although Asian sources are very prominent in the market place in the early twenty-first century.

Granite pegmatite mining has greatly changed. The invention of dynamite, patented in 1867 by Alfred Nobel, radically changed all types of mining. The rapid development of technology to support mining, including using high explosives, further revolutionized mining and, as a consequence, the production of gem stones. Blasting with low explosive materials such as gunpowder began in the year 1600, but the work of mining still depended on human motive power. Bigger vehicles and bigger tools also made many hard rock occurrences such as granite pegmatites easier to work, although blasting powders may have continued to be used in some pegmatite districts into the twentieth century as gem pockets were near the working face of the quarry.

The first interest in granite pegmatites was for mica, but extended to feldspar used in the manufacturing of porcelain plates or sanitary fixtures. Rare-element-bearing minerals such as those containing lithium or tantalum were in high demand during various time periods. At one time, South Dakota was one of the principal regions supplying a late nineteenth century fad for over-the-counter lithium-based beverages and medicines. During the Cold War [1945-1991], granite pegmatites were extensively mined around the world for the beryllium mineral, beryl. However, the mining of the enormous Spor Mountain bertrandite deposit in Utah beginning in the 1970s virtually eliminated beryl's use as an ore. Similarly inexpensive substitutes have been found for some feldspar uses and the current demand for this mineral is now easily supplied by a few granite pegmatites. The largest supply of commercial mica is now from India and few granite pegmatites can match the quantity of mica produced there or the low cost of labor to mine the mica. Granite pegmatites are now primarily mined for rare-element minerals and gems.

The future of mining granite pegmatites is still bright. The lithium battery is promoted as the power source for automobiles and during the writing of this book two new spodumene-rich pegmatite discoveries were announced. The Big Bird pegmatite lithium discovery in the Northwest Territories of Canada, somewhat east of Staples Lake, was formally announced as well as the discovery of some lithium-rich pegmatites in North Carolina.

Granite pegmatites occur in clusters or swarms in a region. Normally, pegmatites may be found enclosed in granite or other "igneous" host rocks or they may be hosted by metamorphic rocks such as schist or gneiss encircling the plutons. As a result, pegmatites are found in areas where there are mountain chains where these rocks have been exposed. Granite masses or plutons may be exposed at the Earth's surface or not, but pegmatite fields without exposed plutons have frequently proven to be associated with plutonic rocks that are still "buried," based on gravity surveys.

There are innumerable pegmatites in the USA, but the economically interesting pegmatite districts have been relatively few. To be "economically interesting," pegmatites have to be large enough to support a reasonable mining venture and have to be close enough to good transportation for the minerals to be brought to market. For minerals such as feldspar or quartz, pegmatites need to be large enough to yield many tons of relatively pure mineral. The source of minerals bought by the ton has to be relatively close or transportation costs become prohibitively high. Gem-bearing pegmatites can be profitable, even in remote mountainous regions where pack animals or humans are the only means of transportation, because a single mineral specimen may be worth several years' salary for a miner.

Granite pegmatites are distributed within discrete districts or "fields. The idea is that a particular pegmatite field may have pegmatites that share a common origin, although there are certainly pegmatite districts with several sources. Exposed plutons may be a kilometer in maximum dimensions or many tens of kilometers across and the associated fields of granite pegmatites are correspondingly large and may be a few kilometers to 100 km long and less than a kilometer or up to 10-20 km wide.

The map of the Staples Lake pegmatite field in the Yellowknife-Beaulieu Region, District of Mackenzie, Canada, shows the relationship of granite pegmatites to other rocks (edited from Kretz 1968). Two named granites are shown in blue that are intruded into schists (shown in green). The granitic plutons do contain numerous granite pegmatites (indicated by "A"), but the map particularly shows granite pegmatites (in red) intruded into the local metamorphic rocks. Almost all of the pegmatites cut across the foliation of the schists (trending just west of north; map edge North-

South). On the scale of the map, the shortest pegmatite shown is about 30 meters long and there are enormous numbers of small pegmatites not shown.

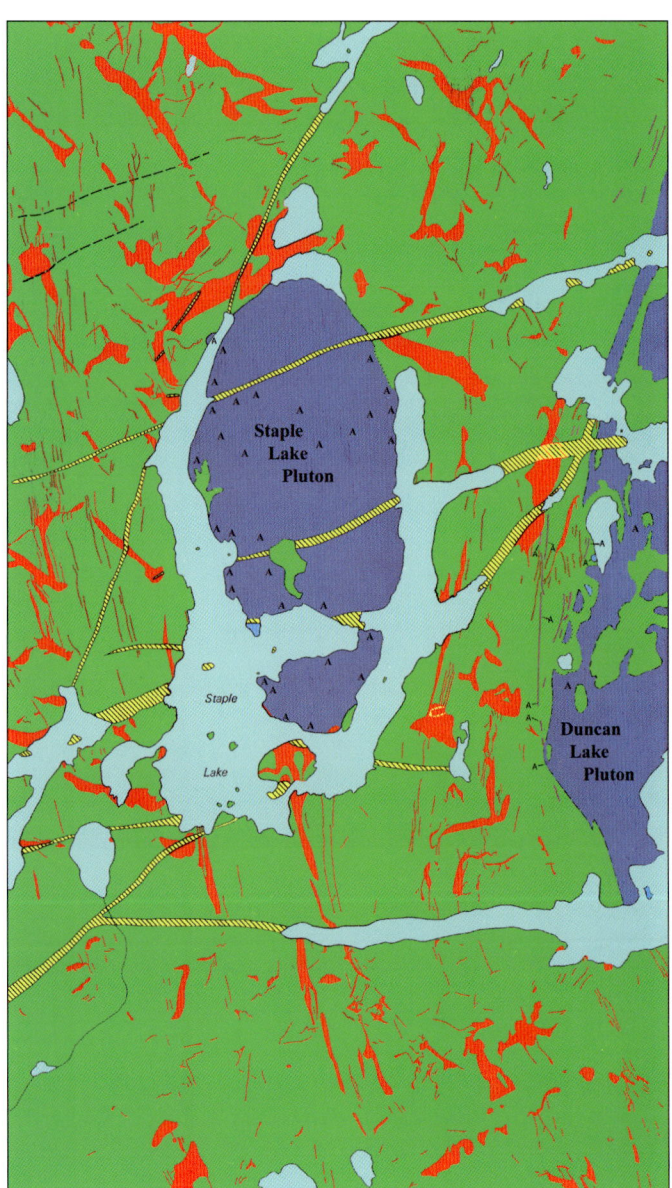

Staples Lake Pegmatite Field, Mackenzie, Canada. Dark blue purple = Granite, Light Blue = Water, Green = Schists, Yellow = Basaltic Dikes, Red = Pegmatite dikes, Light gray purple = Small Granite Veins (middle right), A=Pegmatite Veins in Granite. Map area = 800 x 1370 meters. Smallest pegmatite dike shown is about 30 meters long. (Kretz, 1968).

Pegmatites in the Staples Lake area are far away from markets that would make them economical sources of minerals, such as feldspar or mica, and for this reason no mining has occurred in these bodies. No rare element-bearing minerals have been discovered in these pegmatites, but that may be due to poor exposure of the interior of these pegmatites. The Staples Lake map does show the concentration of pegmatites that may be found in a granite pegmatite field. In areas closer to civilization, granite pegmatites have generally been thoroughly prospected and many discoveries were made in pegmatites that appeared to be "uninteresting" at their surface. One wonders if the Staples Lake granite pegmatites contain vast reserves of spodumene and untapped crystal pockets?

Almost all of the granite pegmatite districts in the USA are inactive with respect to industrial minerals: feldspar, mica, beryl, etc. The notable exception is the Spruce Pine pegmatite field of North Carolina that has produced most of the USA's feldspar for almost a century. Gem mining, however, remains the biggest interest for pegmatite miners in the USA, with perhaps 100 or more quarries and prospects currently operational in California, Colorado, Maine, South Dakota, and Virginia. Additionally, some granite pegmatites are owned by not-for-profit organizations, which administer former commercial mineral locations for educational purposes, and a few dozen small companies maintain fee-access localities for visiting naturalists.

In the USA, granite pegmatites are distributed all along the Appalachian Mountain Range. Important districts are found in the Black Hills of South Dakota, in the Llano Uplift of central Texas, and all along the Rocky Mountains, Sierra Nevada, Cascade, and Coastal Ranges of the western USA and the Alaskan and Brooks Ranges in Alaska. Although, the Hawaiian Islands are igneous in origin, they are dominated by basaltic rocks and are devoid of granite pegmatites. In Canada, granite pegmatites are abundant along the southern and western margins of the Canadian Shield as well as the northern extension of the Rocky Mountains.

Jahns et al. (1952) described in detail the pegmatites of the Appalachian Mountains in the southeastern USA. They mapped pegmatites in 12 different fields and reported on their economic significance. One of the fields, the Amelia in Virginia, has an internationally known gem pocket-bearing pegmatite and there are dozens of internationally known rare-element and rare-mineral pegmatites in their map area.

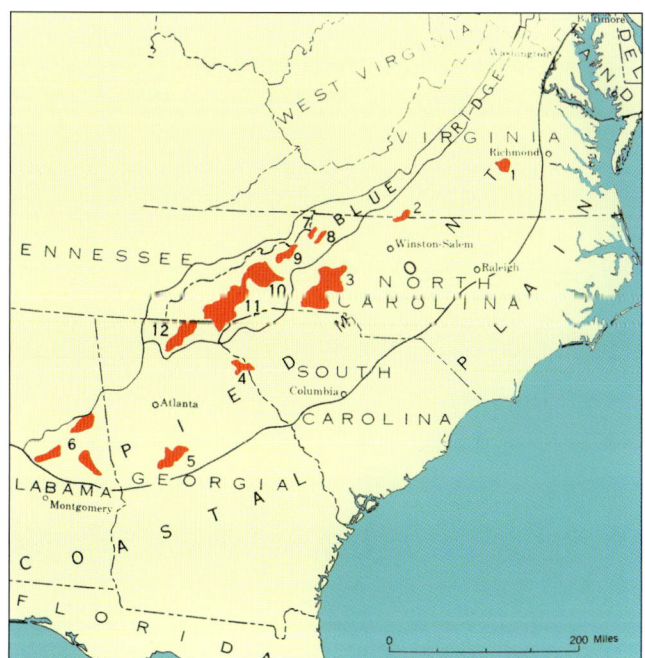

Large Pegmatite Fields or Districts in the Southeastern USA. Piedmont Region: 1 = Amelia, 2 = Ridgeway-Sandy Ridge, 3 = Shelby-Hickory, 4 = Hartwell, 5 = Thomaston-Barnesville, 6 = Alabama, Blue Ridge Region: 7 = Jefferson Boone, 8 = Wilkes, 9 = Spruce Pine, 10 = Buncombe, 11 = Franklin-Sylva, 12 = North Georgia. At least one internationally known gem pocket pegmatite in the map area. From Jahns et al. (1952).

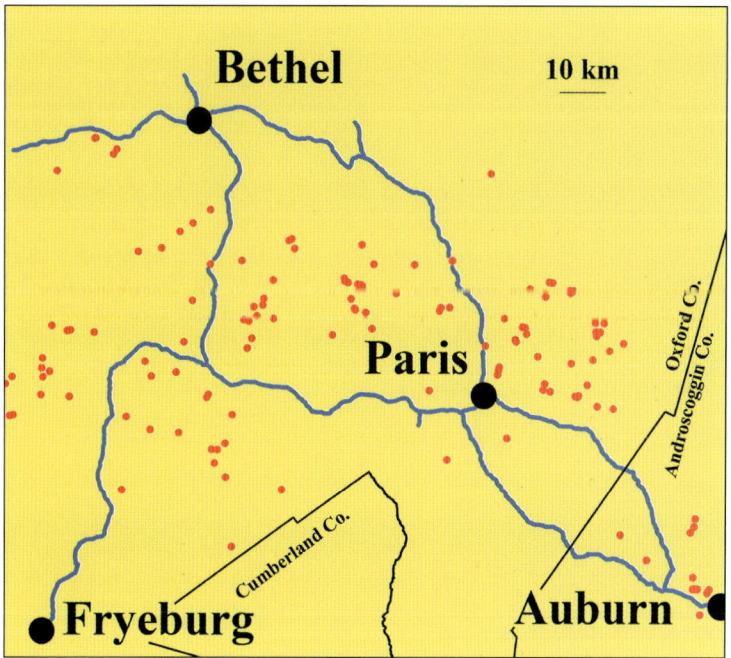

Mined pegmatite locations in western Maine. Pegmatites in the area were mined for feldspar, mica, beryl, and gems. At least three internationally known gem pocket pegmatites in the map area.

The map of Oxford/Androscoggin Counties, Maine, shows several pegmatite fields overlapping an area 50 x 60 km containing about 25 significant gem pocket-producing granite pegmatites, three of which have international significance. The red dots show where granite pegmatites have been mined and there are a great many pegmatites of varying sizes in the region. These pegmatites are distributed in an arc that roughly parallels the outcroppings of the Sebago and Songo Plutons.

Gem pocket-bearing pegmatites are rare in many pegmatite fields. The Chandler Pegmatite in New Hampshire is about the only gem pocket pegmatite in that state, despite having one of the largest concentrations of mined pegmatites in the USA. Connecticut has only a few gem-pocket bearing pegmatites, of which the Gillette in Haddam is the most famous. The pegmatites of Amelia, Virginia, are the only gem producing granite pegmatites in the southern Appalachians. Similarly, South Dakota has an enormous number of mined pegmatites and there is still hope for a major gem discovery. Similarly, Texas, New Mexico, Arizona, Wyoming, and Montana are awaiting recognition, despite having mineralogically rich pegmatites.

In Colorado, Mount Antero's pegmatites are exciting to collect in despite their high elevation (3000+ meters) and the mountain's legendary thunderstorms. Similarly, pegmatites occur in the Lake George and Pikes Peak districts of Colorado, in northern New Hampshire, and the Sawtooth Mountains of Idaho. These districts have thin pegmatite lenses, which are found in normal granite. These pegmatites show abrupt changes of crystallization from the host granite and have open "cookie jar" pockets. These pegmatites generally have a simple suite of minerals, although some gem crystals may be found, particularly topaz and smoky quartz. Blocky microcline crystals and smoky quartz crystals usually are the most important crystals lining a pocket in these districts.

Nothing so far has been said about how pegmatite-forming fluids travel through rock. Perhaps in a very plastic environment bulges of fluid might be forced along foliation in a rock. In a somewhat brittle environment, the fluids may be forced along fractures, although the amount of force needed may not be adequate to push aside two mountains or the Earth to form a vein in between. Rocks under tension may have enough regional forces tearing them apart that the combination of fluid pressure and fracturing would allow fluids to infill a fracture. No matter the method of injection of a fluid, the process of crystallization may be influenced by the host rocks' temperature.

One important factor in crystallization is loss of heat. If the host rocks are about the same temperature as the pegmatite-forming fluid, heat loss from the fluid will be slow, with a consequent slow rate of crystallization. If the host rocks are very much lower in temperature from the intruded fluid, then there may be very rapid heat loss from the fluid to the rocks and crystallization should be relatively rapid. Granite pegmatites begin to crystallize in the range of 600-700° C depending on their composition and environmental factors.

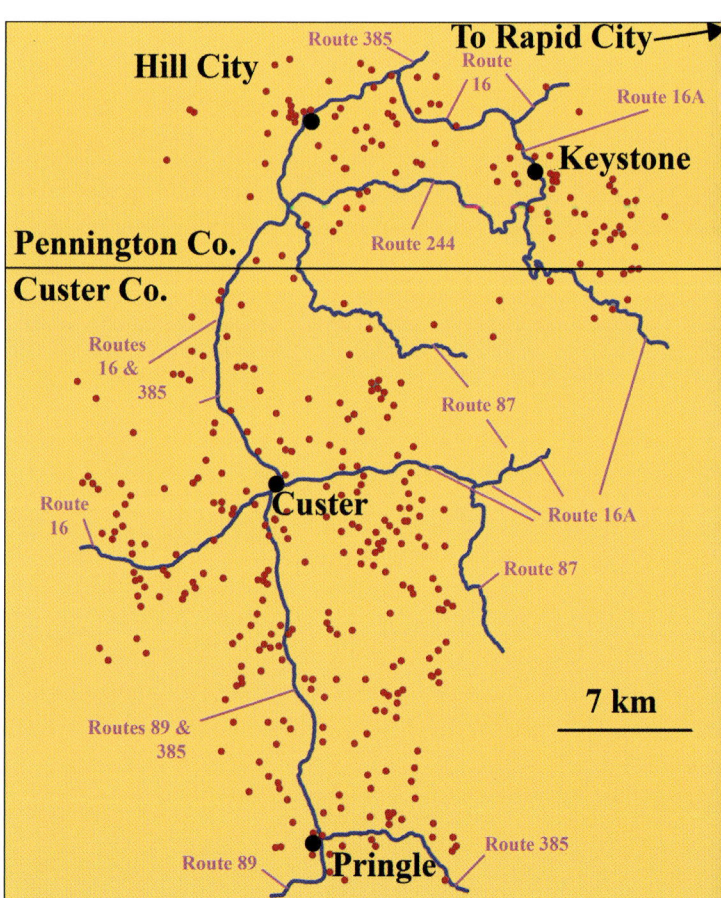

Mined pegmatite locations represent a very small fraction of the pegmatites in any district. Western South Dakota.

Pegmatite Fields or Districts in Southern California, the most important gem producing granite pegmatite area in North America. Five of the eleven districts (named in black; all district boundaries = green; pegmatites in red) in the area account for the great majority of gem production. At least ten of the individual pegmatites have gem pocket occurrences with international significance.

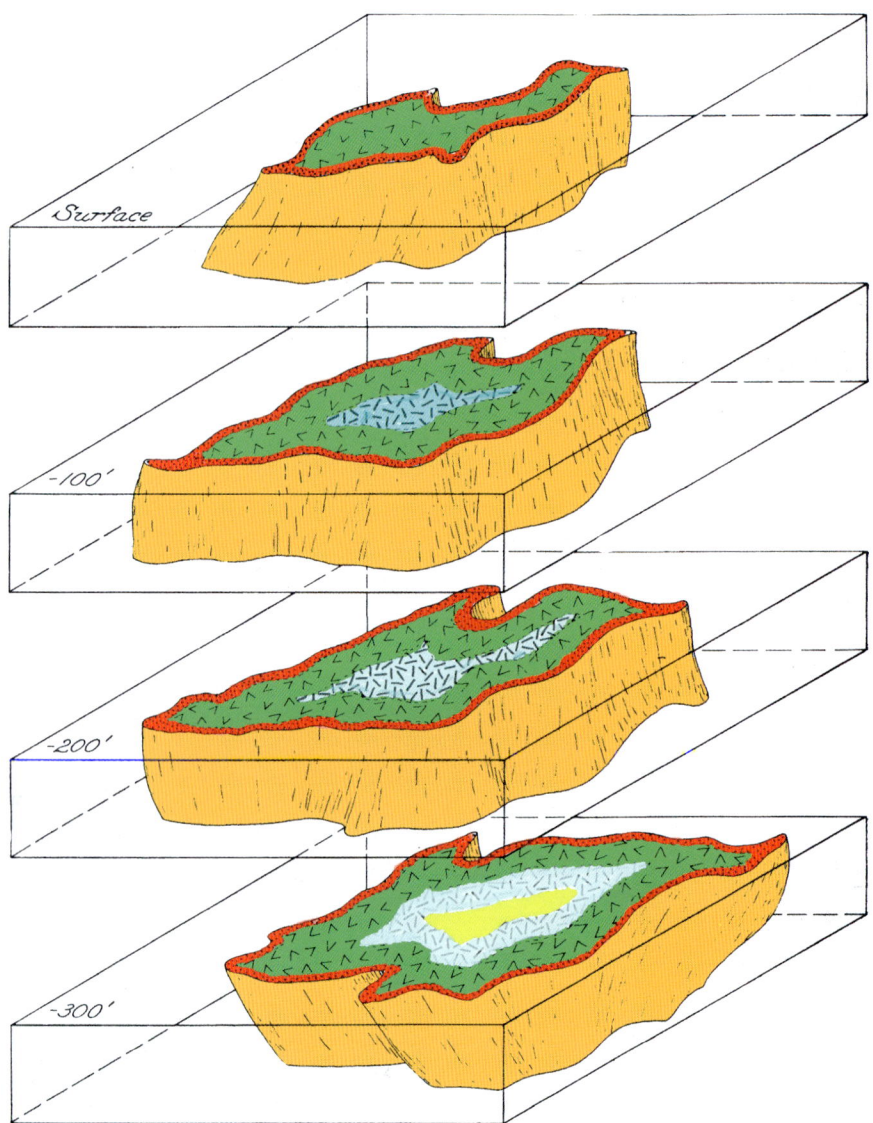

Hypothetical pegmatite showing concentric zones. The various levels show that zones can only be revealed if they are dug into. The surface of a pegmatite may not show interesting minerals. Very deep cuts may reveal additional zones. Redrawn from Cameron et al. (1949).

Chapter Two

The Interior
of Pegmatites – Zones

Granite pegmatites formed from a fluid and crystallized from the outside toward the inside. Because of this universal character, the changes in mineralization in a granite pegmatite are concentric, although they may be asymmetrical. Nonetheless, one of the complications of collecting or of mining minerals and gems in a granite pegmatite is the changeability of the insides of pegmatites. Many of the very small pegmatites have mineral changes starting at the contact of the pegmatite with the host rock and then progress in a series of stages, while some big pegmatites may show mineral and textural changes which may be difficult to discern. On the other hand, there are small pegmatites in northern Pakistan or in San Diego County, California, where the pegmatites are very thin 1-2 meters, but have among the most beautiful crystals known from the world's pegmatites. These thin pegmatites rapidly progress from the contact to a central system of channels and bands of minerals with extraordinary mineral variety.

Granite pegmatite quarries that have been long abandoned may be particularly difficult to explore and understand. Weathering, obscuring plant or lichen growth, staining, water seeping, dirt and rock debris, etc. will obscure the various mineral relationships in the excavation. The evolution of pegmatite fluids is important in crystal pocket formation and the miner and collector need to recognize pegmatite evolution. The original pegmatite-forming fluid had the average chemical composition of a normal granite. In order to have rare-elements in high enough concentration to form gem pockets and gem minerals, changes have

to occur in the proportions of elements in the crystallizing fluids. Zoning is evidence of those changes. By the time a pegmatite actually forms replacement units and crystal pockets, almost all of the original fluid has crystallized. Gems are typically formed in the last fluids remaining. Usually, gem pockets are found in the last 1% of the pegmatite that crystallized. There are some pegmatites that are exceedingly rich, but if gem pocket-bearing granite pegmatites are rare, then rich ones are the rarest of all. Of the millions of granite pegmatites that have been mined, it would be difficult to find more than several hundred that have been noteworthy gem producers.

Oftentimes, large pegmatites have a more or less concentric pattern of mineral variability and the distance from contact to gem pockets may be considerable. Additionally, some pegmatites have noticeably influenced the host rocks they are in and something might be suspected of a pegmatite's contents even before it is dug into. Gem-bearing pegmatites generally have a large variety of minerals, partly because the chemical elements that are essential to form gem minerals are also available to form other minerals. Pegmatites that have gem pocket minerals frequently have more than 40 different species when a detailed list is available. Pegmatites that have been economically important for ore minerals such as feldspar or mica may have at least ten species present. However, some pegmatites may have a large species list and still be barren of gem-crystal pockets.

A real pegmatite showing visible zones. The top medium gray rock is schist. Crystallization and zoning progresses to the interior. The Border Zone is too small to be visible at this viewing distance. The white band of minerals in the top of the rock face is a Wall Zone with a variable thickness to about 1 meter consisting of albite, microcline, and quartz. Below that is the First Intermediate Zone (also called Outer Intermediate Zone, but because of the large number of zones in this pegmatite, numbering is the easiest means of labeling). The First Intermediate Zone is rich in muscovite, giving a gray tone to the rock along with albite, microcline, quartz, and dark triphylite. Maximum thickness is 1 meter in field of view, but increases to over 2 meters in exposures to the right off camera. Below that is the Second Intermediate Zone in the large white part of the supporting pillar consisting of microcline plus quartz. A thin grayish Third Intermediate Zone above the base of the pillar contains quartz, microcline, and spodumene. A lighter gray Fourth Intermediate Zone exposed on the base of the pillar contains muscovite, quartz, microcline, and cleavelandite. According to the geologic map, behind the pillar there is a Core Zone containing quartz with perthite, muscovite, and cleavelandite. FOV = 15 x 25 meters. Main Pegmatite, Newry, Maine. See associated cross-section maps.

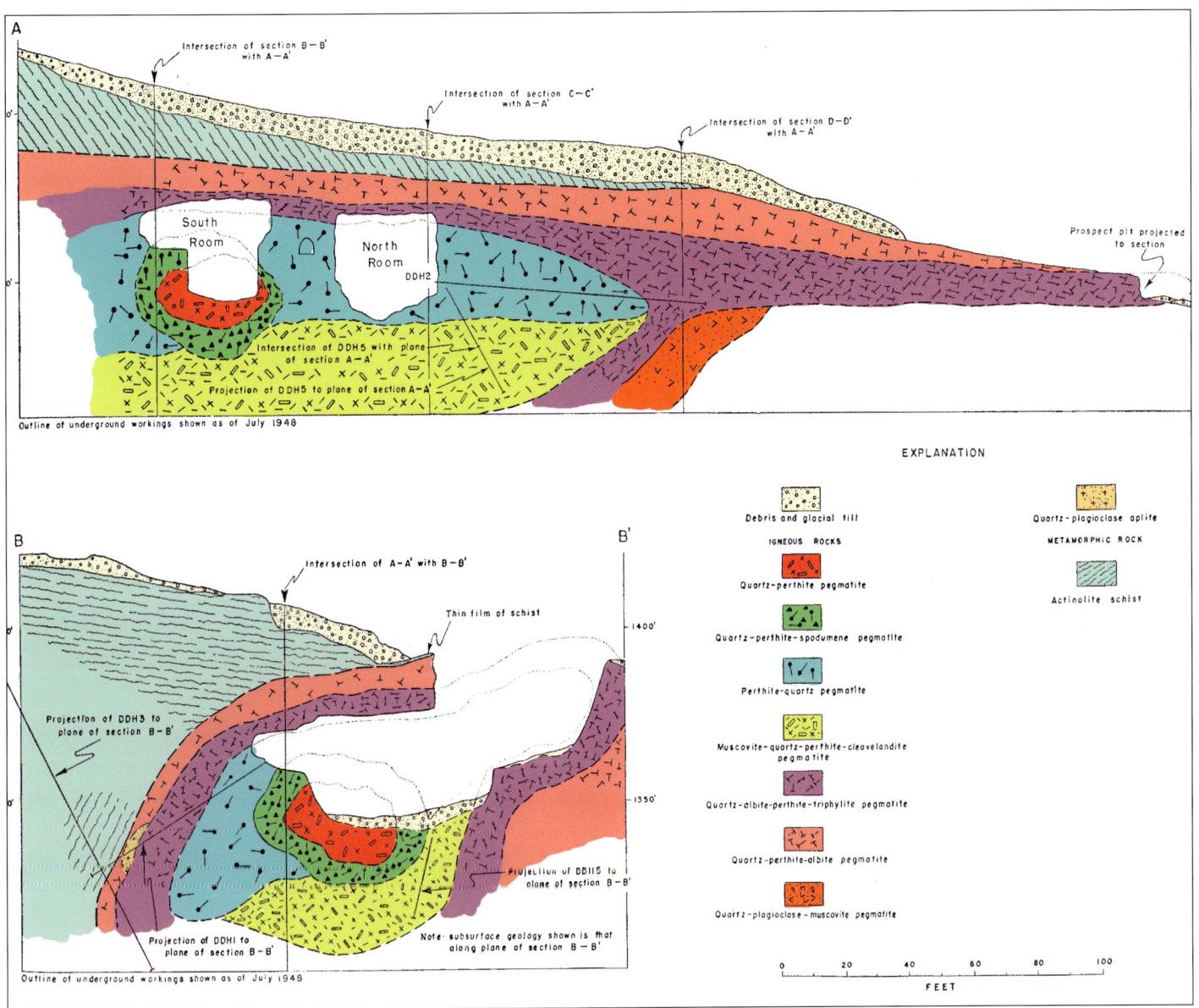

Geologic Map of the Main Pegmatite, Newry showing the boundaries of the zones. Top layer in pale yellow is dirt. Mint green represents the schist host rock. Light red brown is the Wall Zone. There is a small medium tan aplite body near the contact on the left of the lower diagram. The purple indicates the First Intermediate Zone. The pale blue and bright yellow zones are also Intermediate Zones and for convenience, the blue is called the Second Intermediate Zone and the yellow is called the Fourth Intermediate Zone. The yellow green is the Third Intermediate Zone and is visible in the preceding photograph, and the red "quartz perthite" pegmatite is probably a Core Zone. A quartz albite muscovite pegmatite appears to underlay the entire well-zoned apex of the pegmatite to an additional depth of 84 meters below the zoned portion of this pegmatite (Shainin and Dellwig, 1955; Barton and Goldsmith, 1968; King, 1980). The underlying pegmatite is fine- to medium-grained (1-4 cm) and the feldspar is mostly albite, but there is localized development of microcline. The underlying pegmatite is equivalent to the Wall Zone and has numerous aplite layers. The maximum thickness of the pegmatite is about 90 meters.

Elbaite requires lithium in its formation, but there are other minerals that also require lithium and which usually form before elbaite does. Spodumene, lepidolite, montebrasite, and/or triphylite are examples of lithium-bearing minerals that may precede gem tourmaline formation. Even gem green and red tourmalines are almost always preceded by black tourmaline. In fact, a large quantity of black tourmaline in a granite pegmatite is an excellent sign and this fact will be often repeated for emphasis. Successful gem mining requires that the miner is able to recognize many different minerals, their proportions in zones, and critical mineral associations. The miner also has to be able to realize when to abandon the workings, because there is little hope remaining for recovering minerals economically. Nonetheless, miners do return to abandoned diggings because they hope that the previous miners missed something.

Pegmatite-forming fluids have generally moved some distance from where they were generated. For simplification, the origin of these fluids may be left to more technical discussions. However far pegmatite-forming fluids have traveled, they had to collect together to form a rock. The granite pegmatites formed at least several kilometers to perhaps six kilometers below the surface of the Earth and at this depth solid materials are under enough pressure so that they behave plastically. Solid rock may "flow" or be deformed under these pressures. Similarly, fluids may be forced from high pressure to lower pressure and form a chamber where they may crystallize. The mineral zones that form in a granite pegmatite are (from outside to inside): Border Zone, Wall Zone, Intermediate Zone, and Core Zone. Very complex pegmatites may contain several Intermediate Zones. Superimposed on the sequence may be a "Pocket Zone." This latter "zone" is not the same as the other zones first mentioned. The Border through Core Zones form by crystallization directly from a fluid and are said to be "structurally controlled." In essence, true zones resemble the overall shape of the pegmatite, although the zones may be unusually or asymmetrically developed. A Pocket Zone is superimposed on the pre-existing minerals and develops from the remnant fluids in the pegmatite body that had yet to crystallize. These fluids are very corrosive with respect to the solid pegmatite and may dissolve some of the earlier-formed crystals. Pocket Zones form at the expense of earlier formed minerals and pockets and their surrounding minerals may cut across the original zones that had formed. The dissolution and replacement process may create cavities and very late crystals may form inside of these openings, creating beautiful minerals.

The erratic distribution of replacement means that a miner may encounter pockets in somewhat unexpected places. In practice, replacement units do not extend very far from the Core Zone and rarely cut beyond any Intermediate Zones. There is little reason to expect that each attempt to excavate a granite pegmatite would yield a gem pocket, but one adage that is important states, "You've got to move rock, if you hope to find tourmaline!" (Frank Perham, personal communication, 1974).

Fracture fillings may also occur in pegmatites. The presence of fluids that may replace earlier-formed pegmatite are enriched in water and other volatiles and may exert a high "steam" pressure. As rocks cool, they contract and the remaining fluids under pressure may follow cracks that form and may deposit unusual minerals along these fractures, sometimes depositing gem minerals. Fractures are usually too small to be an important source, however.

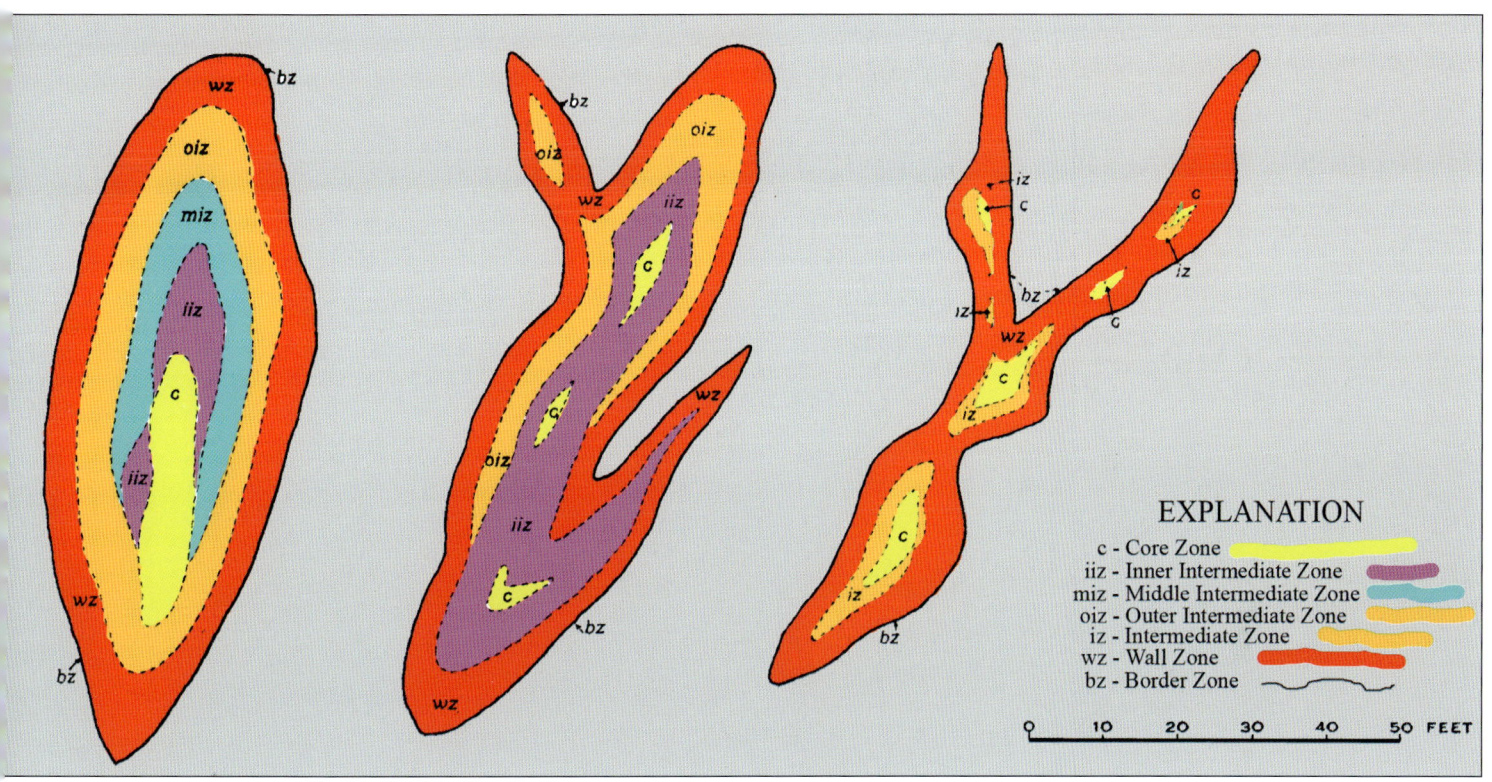

Hypothetical pegmatites showing concentric zones. Left drawing shows a concentrically zoned pegmatite with Core and two Intermediate Zones asymmetrically developed toward the Earth surface. Middle shows an irregular pegmatite with a somewhat symmetrical development of inner zones, but with one isolated zone. Right shows an irregular pegmatite with five isolated Core Zones and six isolated Intermediate Zones. Redrawn from Cameron et al. (1949).

The discussions about zones may give the impression that zones are easy to spot and they are regular, symmetrical, etc. Zones are particularly difficult to recognize when the exposed face of a pegmatite is weathered, has vegetation growing on the surface, or if there are weathering stains. Zones are more apparent in fresh exposure, but still may have too much variability and it may be difficult to trace the boundaries of the zones. Some minerals are distributed in clusters or may not be randomly spaced. If anything, irregular distribution of minerals is characteristic of pegmatites, although it may be possible to discern an overall "structure" to the distribution of minerals. Because the broad spectrum of grain sizes is part of the definition of pegmatite, it is expected that mineral grain sizes within a zone will vary widely. There may be a systematic gradation in grain size across a zone. The variations do not make understanding a pegmatite hopeless; the observer just has to be aware of the complications when simplifications encounter reality.

Another problem associated with pegmatite mining is knowing where the current Earth surface exposes a pegmatite. The idealized drawings do not prepare the prospector for knowing where they are when viewing an exposed surface. The idealized drawings should never encourage a miner to continue digging in the vain hope that mineral assemblages will surely improve if only they continue digging until they encounter a mystical pocket zone that may reveal itself in the next few meters of mining.

Even In a well-known gem pocket producing pegmatite field, very few of the pegmatites have gems in them. Some authorities suggest that 1% of the "reasonably-sized" pegmatites have gem pockets at all and very few of those pegmatites yield gem tourmaline, beryl, or spodumene. In a pegmatite field where there has been extensive mining, there has already been a selection process by the initial prospectors. Prospectors for industrial minerals have rejected pegmatites with too many impurity minerals because they would not have "clean" feldspar, sheet mica, or other minerals that were easy to mine and sell. Many would suggest that the "rejected" pegmatites would be the most interesting to explore.

In any discussion of zones, it is interesting to consider some of the common collectible minerals that occur in the different parts of a pegmatite. With over 800 granite pegmatite mineral species to choose from, the illustrations cannot come close to doing justice to representing species diversity. A few selected minerals will be shown for each zone, but the gallery items will be skewed toward the mineral collector's interests. Probably, some of the specimens in the illustrations have actually come from an adjacent zone than they have been attributed, but are included for descriptive purposes. As most of the specimens illustrated are also from the marketplace, the reader's indulgence is requested. Nonetheless, the images should help reinforce the idea that the mineralization of zones evolves from the outside of the granite pegmatite towards the interior. The real test will come when granite pegmatites are examined in person.

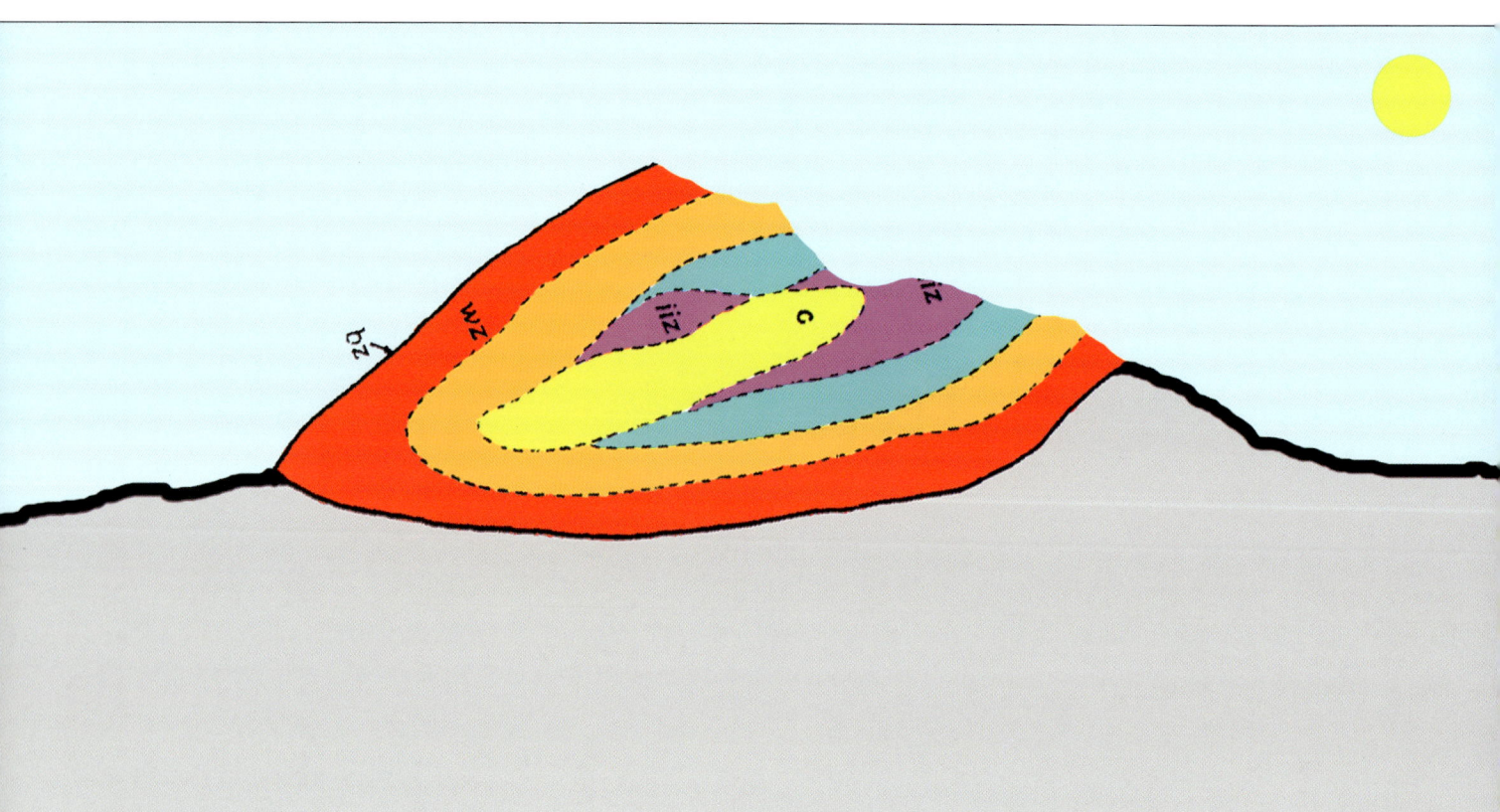

Hypothetical pegmatite exposed as a hill crest. The landowners on the left side may look at the outcrop and see only Wall Zone minerals and conclude their pegmatite is worthless. The landowners on the right may look at an outcrop with gem pockets exposed on the surface.

The Border Zone

For crystallization to begin, heat needs to dissipate from the fluid into the surrounding rocks and with loss of heat, crystals begin to form. If the fluid is injected into rocks that are significantly cooler than itself, heat will be conducted away from the fluid rapidly and small crystals will form on the border of the fluid chamber. The lining of small crystals at this contact forms a "Border Zone." The minerals in a border zone are usually representative of the original fluid and there may be feldspar (usually microcline and/or albite/oligoclase), quartz, and very minor amounts of mica (biotite and/or muscovite). Even in large pegmatites many tens of meters in maximum dimension, the border zone may be only a few centimeters thick and in most cases, there are no interesting minerals in the Border Zone. When pegmatite-forming fluids were injected into relatively hot rocks and/or the pegmatite fluid is rich in volatiles, a Border Zone may be poorly developed.

Because the Border Zone forms an insulating layer around the pegmatite-forming fluid's chamber, granite pegmatites might be thought to be closed off from further interaction with the host rocks, but there are probably many influences still at work. Boron is fairly volatile and there are compounds of it that are vaporous and can migrate into the country rock and may form reaction minerals surrounding a granite pegmatite. Adjacent limestone or marble units may develop minerals such as vesuvianite, grossular, diopside, epidote, etc. Boron-bearing fluids may form tourmalines outside of the pegmatite. Volcanic rocks may have micas developed on the outside contact. Contact mineralization should not be ignored when exploring for just the right granite pegmatite to mine.

Granite pegmatite (right) in contact with granite (left side). There is a white band of Border Zone pegmatite in contact with the granite. The right side shows coarser grained pegmatite than the border zone. New Gloucester, Maine.

Light-colored Granite pegmatite in contact with gray green schist with irregular band of Border Zone outlined by iron-stain. The upper side of the Border Zone shows a large mass of feldspar. Note black tourmaline along contact where pegmatite fluid reacted with schist. Coin = 1.8 cm. Dunton Pegmatite, Maine.

Well-developed black tourmaline in contact schist. 20 x 12 cm. Dunton Pegmatite, Maine.

The Wall Zone

Most of the bulk volume of granite pegmatites is composed of Wall Zone minerals and textures. Wall Zones are mostly composed of common minerals such as: feldspar (microcline and/or albite), quartz, and mica (biotite and/or muscovite) with occasional garnet, and black tourmaline, but apatite, beryl, and even columbite may be present. In pegmatites with important amounts of Rare Earth Elements, minerals such as monazite may be conspicuous, but many Wall Zone minerals that first appear in that zone are also found in the rest of the zones deeper in the pegmatite. As the common minerals, they select only the nutrients they need to crystallize; the remaining fluid of the pegmatite becomes enriched in the components that are not selected. Depending on the actual chemical composition of the initial fluid and the volume of the fluid, a granite pegmatite may crystallize with relatively few species. This is the case with 99+% of granite pegmatites. Crystallization does not lead to gem pocket formation, etc. There are complexities superimposed on the various crystallization paths that a fluid may take, but most pegmatites are "simple." They have few poorly developed zones with uninteresting minerals. The following discussion presumes a fluid with favorable conditions (composition, crystallization sequence, etc.), which leads to a complex pegmatite with well-developed zones, and, of course, crystal pockets.

Even in the Wall Zone, there are minerals which may be used as predictors of "future" minerals which may be present or absent. Black mica, biotite, is not necessarily a bad sign, but black micas are rich in iron and iron-rich mineral associations do not frequently indicate gem pockets nearby. However, there are thin and thick pegmatites that have black mica near the contact. Black mica may be found in contact with the common white mica, muscovite, but deeper into a granite pegmatite, muscovite is usually the important mica. Tourmaline miners welcome the early appearance of black tourmaline, usually schorl, in a granite pegmatite. Boron is important in all tourmalines and black tourmaline will form as long as there is abundant iron. Tourmaline itself may be a mineral responsible for lowering the amount of iron in the residual fluids in a granite pegmatite. The beautiful gem tourmalines are low in iron, but otherwise share much in common with black tourmaline.

The pegmatite is in contact with a schist in upper left. The diagonal Border Zone is several centimeters thick and the grain size of the pegmatite dramatically increases including rod-like crystals of black tourmaline. FOV = 1.5 x 2 meters. Emmons Pegmatite, Maine.

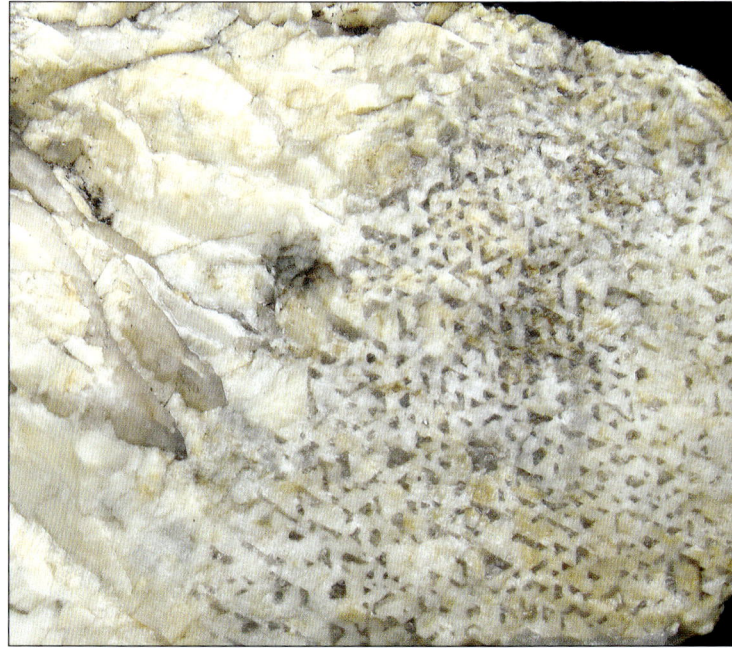

Graphic granite may be a common feature of Wall Zones. FOV = 7 x 7 cm. Keith Pegmatite, Maine.

Minerals That May Be Seen in Crystals in the Wall Zone

Micas are universal in granite pegmatites and frequently mica crystals may express a characteristic shape. Black annite crystal surrounded by clear muscovite crystal. 5 x7 cm. Wheeler Pegmatite, Maine.

Almandine garnet in microcline and muscovite. 8 x 9 cm. Hedgehog Hill Pegmatite, Maine.

Black tourmaline in quartz. 12 x 8 cm. Barnett Pegmatite, Canton, Maine (*Norman Davis Collection*).

Beryl in quartz. 5 x 7 cm. Bumpus Pegmatite, Maine.

Brown Monazite-(Ce) in quartz. 1 x 1 cm. Zlatoust Mountains, Russia.

Intermediate Zones

Most pegmatites have no Intermediate Zones. Intermediate Zones are merely succeeding zones between the Wall Zone and Core Zone, but the "action" may begin to appear in Intermediate Zones. The changing granite pegmatite-forming fluid expresses itself in a new set of minerals or ratio of various common minerals and grain size may be dramatically bigger compared to the Wall Zone. Giant crystals are more likely to be present in the Intermediate Zone than in the Wall Zone. Giant beryl crystals or giant spodumene crystals may be found in them and there are excellent examples of giant crystal formation in New Mexico, South Dakota, Maine, and North Carolina pegmatites. If feldspar crystals without sharp crystal boundaries may be included,

then almost all of the so-called complex pegmatites have giant crystals in addition to many of the simple pegmatites. Micas may be abundant and coarse-enough to be harvested. Feldspars may be valuable in a district where there is a grinding mill and rare-element minerals may be abundant enough to be sold for a profit. Despite the valuable minerals, gem pockets are not usually present in Intermediate Zones, but the presence of such a zone is evidence that changes occurred in the process of the pegmatite's formation. Conversely, there are quite a few granite pegmatites that have an Intermediate Zone, but there are no gem pockets or particularly rare minerals. Nonetheless, a "complex" pegmatite may also contain a bonanza of gems should it really get complex. A pegmatite is only complex in that it has more than the standard internal pegmatitic texture.

Beryl and black tourmaline may persist in occurrence into an Intermediate Zone, frequently increasing in abundance. Upper view: Beryl, black tourmaline, muscovite, quartz, and albite. FOV = 50 x 65 cm.

Phosphate pods may first appear in the Intermediate Zone. FOV = 1 x 0.7 meters. Purplish black heterosite replacement of original triphylite in quartz, muscovite, and albite. FOV = 15 x 20 cm Mount Marie Pegmatite, Maine.

An Intermediate Zone may form when water is dramatically increased over its proportion of the original fluid and rare-elements in the form of rare-element minerals may be very conspicuous as they, too, have been concentrated in the late fluids. The process of concentration will be mentioned many times because the leftover concentrated or enriched elements are the stars of the granite pegmatite "show."

A few pegmatites have three or four Intermediate Zones. There is nothing that implies that Intermediate Zones possess particularly rare minerals, although if a granite pegmatite does possess rarer minerals, they usually express themselves in Intermediate Zones rather than in the Wall Zones. As pegmatites crystallize, the common elements get used to form the common minerals, feldspar, mica, quartz, garnet, black tourmaline, etc. The elements that form gem minerals may be almost the same as the minerals that are removed early. For example, topaz does not really have much in the way of rare elements except fluorine, yet it is reasonably rare. Much has been made of the role of lithium in making desirable tourmaline and its frequently associated minerals: lepidolite, spodumene, montebrasite, and triphylite. Lithium is not particularly removed from the pegmatite-forming fluid by the common minerals and lithium may become enriched relative to its starting concentration. When these lithium-bearing minerals form, they begin to remove large amounts of lithium and the miner hopes that they have not removed too much lithium. The granite pegmatites in South Dakota have large amounts of triphylite, sometimes spodumene, and even montebrasite, but almost no elbaite. The world famous Palermo #1 Pegmatite in New Hampshire has masses of triphylite that may be up to 5 meters long. Despite the concentration of lithium required to make large masses of triphylite, there has been no discovery of gem pockets full of lithium tourmaline at the Palermo #1 Pegmatite as it is poor in an essential ingredient of tourmaline: boron. Similarly, large masses of the lithium aluminum phosphate hydroxide, montebrasite, were mined by the ton at the Tin Mountain Pegmatite in South Dakota, but relatively little elbaite was discovered. It is a curious fact that the Black Hills granite pegmatites in the Keystone-Hill City-Custer-Pringle District are very rich in triphylite and despite the lithium present to form this mineral, the Blacks Hills have virtually no gem tourmaline, despite there sometimes being a great deal of black tourmaline in the deposits. In most cases, boron is probably very low in the lithium-rich Black Hills pegmatites.

Although gem spodumene is very much prized, crystals of this mineral rarely have any gem potential when encased in quartz or other pegmatite minerals. Spodumene. 12 x 17 cm. Harding Pegmatite, New Mexico.

Spodumene. 11.5 cm long. Strickland Pegmatite, Connecticut. *Photo courtesy Russ Behnke.*

In North Carolina, there are several spodumene-bearing pegmatites, which also have no gem tourmaline, although there was plenty of lithium to make spodumene. Gem miner Frank Perham has the axiom, "Spodumene next to a pocket is the kiss of death for gem tourmaline." Just as astronomers have a "Goldilocks Zone," where everything is just right for life, so do gem miners where everything is just right for pockets.

Minerals begin to form when the concentration of their components becomes high enough. Beryl may begin to form in the Wall Zone or continue into an Intermediate Zone. In fact, beryl may be present in almost every zone in a particular pegmatite. The miner looks for mineral changes and evaluates the mineral assemblages. Mineral diversity is important and hardly a gem pocket granite pegmatite has fewer than 40 different species after a careful examination.

There are many accessory minerals in Intermediate Zones. Minerals that have had commercial importance include the columbite/tantalite group. Columbite when pure is dominantly niobium plus iron and has very low value. The common species is called columbite-(Fe). Columbite/tantalite group minerals are sensitive to chemical changes that have occurred in a pegmatite-forming fluid and niobium may be depleted in late fluids and a related element, tantalum may be enriched. In contrast to columbite, tantalite is very valuable as tantalum is in very high demand to make specialized alloys. Interestingly, the iron-rich member of the tantalite series is very rare and the common end of this series is manganese-rich. Ordinarily, it is difficult to see the probable chemical composition of a mineral, but iron-rich members are black, while iron-deficient, very manganese-rich members are brown to red, as least as a crushed powder. The miner is interested in the extremes of crystallization, so very manganese-rich tantalite is easy to spot as it has red internal glints to actual red body color. Intermediate compositions are difficult to detect, but density is a good tool to use in knowing the tantalum contents.

Almandine. 6 x 6 cm. Mount Mica Pegmatite, Maine.

Arsenopyrite. 8 x 10 cm. LaFlamme Pegmatite, Maine.

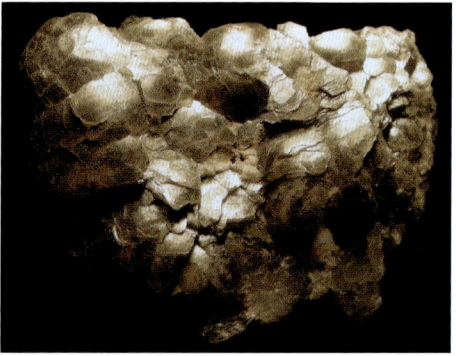

"Ball" Muscovite. 9 x 16 cm. Lord Hill Pegmatite, Maine.

Almandine. 8.9 cm wide. Russell, Massachusetts. *Photo courtesy Russ Behnke.*

The elements iron and manganese are also important indicators of changing fluid composition. If iron makes black tourmaline, it is obvious that a miner wants to find crystal pockets where iron is low. Fortunately, iron can rust (oxidize) and change from a commonly available form (Fe^{2+}) to one that fits in other minerals (Fe^{3+}). Manganese is resistant to oxidation and it is responsible for many or the red and pink colors of late stage minerals. Miners particularly appreciate having lithiophilite in their pegmatites as it is already manganese-rich compared with triphylite. But, despite having abundant lithiophilite, the Fillow Pegmatite in Branchville, Connecticut, did not produce many crystal pockets and no rubellite.

Sulfides, and rarely arsenides, may form in small pods in Intermediate Zones. Most of the sulfide pods are iron-rich and the miner is not encouraged by their presence, except as a way for crystallization to remove iron from the original pegmatite-forming fluid.

Columbite-(Fe). 7 x 8 cm. Showing divergent black blades in red microcline, Quadeville, Canada.

The Core Zone

Cores Zones may be fairly unspectacular or they may have world record sized crystals along their margins with a Wall Zone or an Intermediate Zone. Giant textures may include feldspars, micas, beryls, spodumene, phosphate pods, etc. One mineral overwhelmingly dominates the Core Zone itself: quartz. The zone may contain only quartz with no pockets or crystals. At present, there is much debate of how such a large mass of quartz would form. The answer goes beyond any idea that there were not enough other elements remaining in the pegmatite fluid to combine with the silica and make more complex minerals. There may be a few actual quartz pockets in a Core Zone, which are where the quartz crystallized without growing against other pieces of quartz, but such primary pockets are uncommon.

One of the peculiarities of quartz cores is that sometimes they may contain substantial quantities of rose quartz. Rose quartz masses owe their color to ultramicroscopic inclusions of dumortierite, a pale pinkish purple mineral that frequently occurs as fibers. The presence of these fibers is what gives rose quartz a property called asterism. Some rose quartz specimens, particularly if they are polished, will show internal reflections making a six-rayed star when viewed with a strong point source of light. Rose quartz crystals are extremely rare, but owe their coloration due to phosphorus instead of inclusions.

Quartz cores may be a dominant central feature of a zoned granite pegmatite such as at Hagendorf-Süd, Germany or at the Palermo #1 Pegmatite, New Hampshire or may be widely spaced as in some pegmatites in Namibia (Witkop, Karibib, etc.), Colorado (Eight Mile Park District, Quartz Creek District), and many other districts.

Quartz (5 x 7 cm) from core zone of Parker Mountain Pegmatite, New Hampshire.

Microcline crystal (1 meter long) encased in light gray core zone quartz. Albany Rose Pegmatite, Maine.

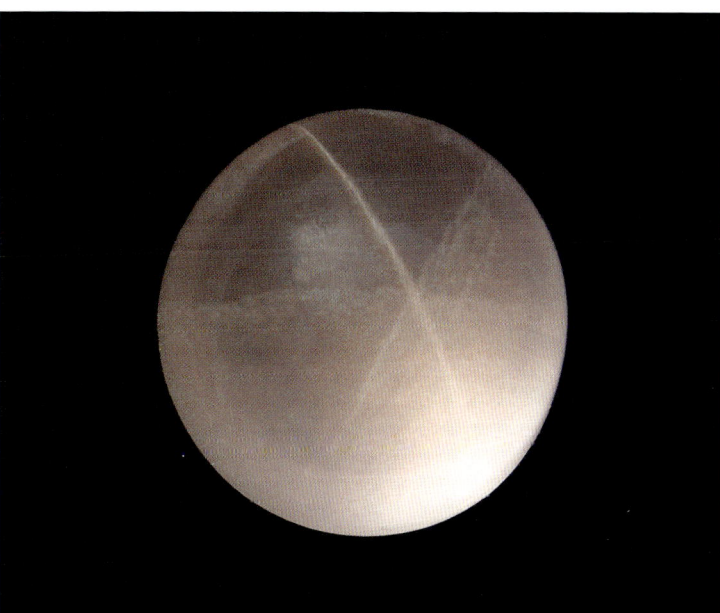

Star rose quartz. 2 cm diameter. Araçuaí, Brazil. *Bill Damron collection.*

Rose Quartz. 4 x 7 cm. Bumpus
Pegmatite, Maine.

Tricia Perham with giant sized microcline crystal (122 x 61 x 54 cm).
Albany Rose Pegmatite, Maine. *Photo courtesy Frank Perham.*

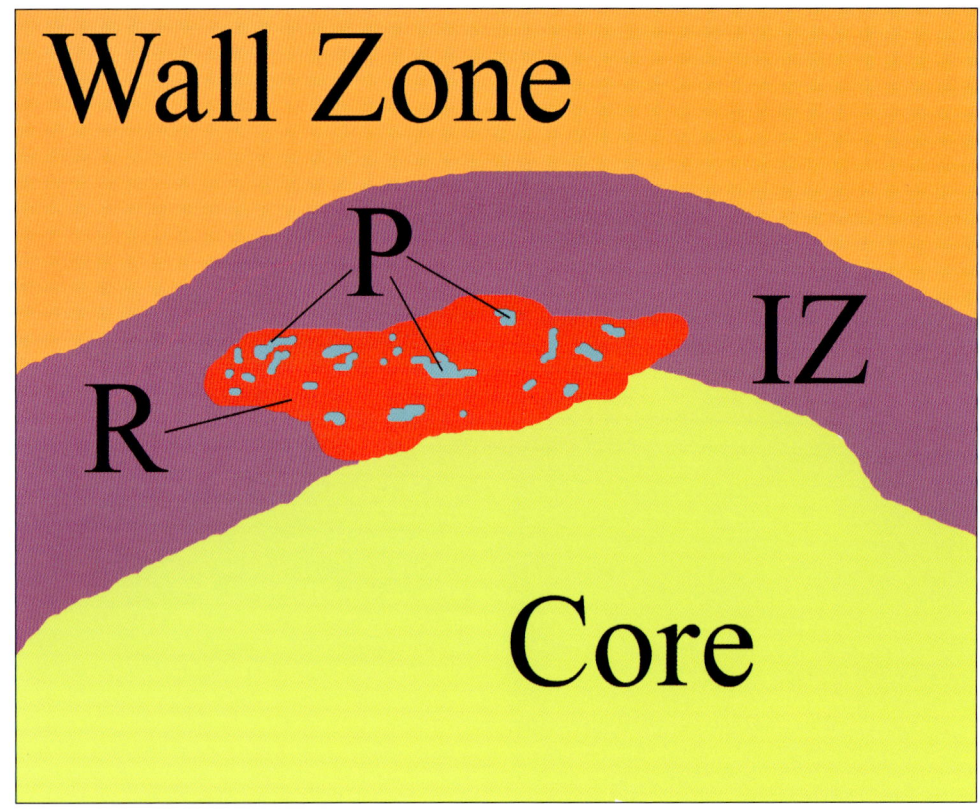

Pockets are replacements of pre-existing pegmatite. Pockets are composed of replacement minerals forming a shell around central cavities where crystals may have grown unimpeded by other crystals. Pockets are usually filled with sandy to clayey minerals. Replacement Zone "R" (red) cutting across primary zones (dark purple) with Pockets "P" (light rose purple).

L-R: Luther Kimball Stone and Loren B. Merrill at Mount Mica Pegmatite, Maine, next to sticks locating exposed crystal pockets (from Bastin, 1911). Photo ~1890s.

Chapter Three

Replacement Units, Crystal Pockets, and Pocket Minerals

Replacement units are not primary features of granite pegmatite and are not true zones. Because pockets are found in replacements of pre-existing pegmatite, they, too, are secondary features. The late portion of the pegmatite fluid to crystallize had undergone many subtractions of common elements, which resulted in water vapor accumulating in the residual fluid as well as rare elements, which also did not "fit" into the common minerals. Some rare elements such as lithium began to appear in minerals in the intermediate zones and continued to produce minerals during replacement mineralization. Replacement Zones suggest to many people that there are also pockets present. Sometimes this is true and Pocket Zone and Replacement Zone are used as synonyms by hopeful miners. The term Pocket Zone will be used here to suggest that there are pockets in the Replacement Zone. However, despite the mechanism for concentrating rare elements, there are many pockets and pocket zones that do not yield gem minerals or particularly rare minerals and some pocket zones are "barren" of valuable minerals or crystals and in some locations even the discovery of simple quartz crystals is considered remarkable. The size and shape of pockets is also highly variable. A pocket may be only a few centimeters across and still provide a few wonderful gem minerals while pockets many meters in minimum dimensions may be a waste of time to clear out.

Entrance of pocket (center dark area) that was 8.5 x 3 x 2.5 meters in Bennett Pegmatite, Maine, in 1924 (Landes Ph. D. Thesis, 1925).

L-R: Ray Woodman and Woodrow Thompson discussing pockets exposed at the base of the quarry wall, 1993, Bennett Pegmatite.

Cleavelandite replacement body cutting across pre-existing cleavelandite. Large cavity to the right was a tunnel leading to gem tourmaline pockets. Dunton Pegmatite, Maine. Shovel = ~0.8 meter (King, 1980).

The zone of pocket formation is three dimensional, but that fact is commonly forgotten. Formerly, pegmatite geologists would describe the replacement and pocket features as a "hood" over the core. The open cavities would be a small percentage of the volume of the pocket zone and would be sporadically distributed through the zone. Whenever a trench is cut through a pocket zone, it could appear that the pockets formed a straight line, but of course any pockets in the center of the trench would be excavated and the pockets exposed along the two dimensional wall of the trench would give the false suggestion that pockets were linear features. Bastin (1911) published a photograph of two pegmatite miners who kept track of gem pockets they had discovered by placing sticks in them as a means of being able to predict any pattern. Pockets may seem to pinch and swell from one to another. A rich small pocket may appear to be an offshoot from a larger chamber. In fact, many pegmatite miners have found some of their best gems in relatively small pockets while some huge pockets may be rich in quartz crystals with only a few or none of the prized minerals.

Diagramming the Pocket Zone

So far, the various diagrams have been drawn to give room for the symbols labeling the details, but one must be aware that an idealized drawing may lead someone to believe that the illustration is always true.

With respect to where zones form in a pegmatite, the shape of the walls of the pegmatite chamber may be very influential. One of the early illustrations in this book showed pegmatites cutting through a granite curbstone and the walls of the dike were not straight but showed a kinked off-set. The same kind of abrupt change may be found in larger granite pegmatite dikes. Pegmatite-forming fluids are probably very viscous, initially. They crystallize from the walls of the vein toward the interior. As the various minerals are removed, which select common elements and leave behind the other elements, the remaining fluid changes composition through this selection process. Water is also left behind in the general process of crystallization and

also exerts its influence on the pegmatite-forming fluid. As water increases, the pegmatite-fluid becomes less viscous and less dense. The changes in the physical properties of the fluid allow the water-rich, rare-element-rich fluid to rise in the dike or chamber: a process involving several stages of fluid evolution.

Extreme rarity where intermediate zones and a pocket zone are continuous along the length of the pegmatite dike.

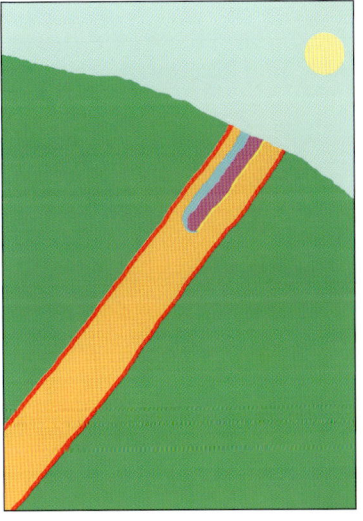

Reality where interesting mineralized zones may have a short range.

Pegmatite dike in curbstone (15 cm thick) rotated to suggest that the kinked offset could act as a trap where rising water-rich fluids might accumulate and where rare-element enriched zones and replacements might occur.

Usually, the pegmatite dike is not much moved in orientation from when it was first emplaced and because of this original position, the effects of gravity may be seen. The less dense fluids may accumulate along the upper side (so-called hanging wall) of the pegmatite giving the pattern of mineralization an asymmetric distribution. The best mineralization in a pegmatite may occur where there is a physical change of shape of the dike. An overhang is an obvious example when the diagram is uncomplicated. Kinks may not be obvious and the outside walls of the pegmatite give a clue to how fluids may have accumulated. The diagrams (this page) show where fluids were trapped under a simple offset fracture and where fluids were trapped under a "roll" of schist.

Pegmatite dike with a sharp kink offset. Rising fluids may collect under the change of slope. The overhang of the dike wall acts as a canopy preventing further rise and is where the "best" mineralization may occur. Blue = Intermediate Zone, Yellow = Core Zone, Dark Purple = Replacement Zone, Light Purple Pockets.

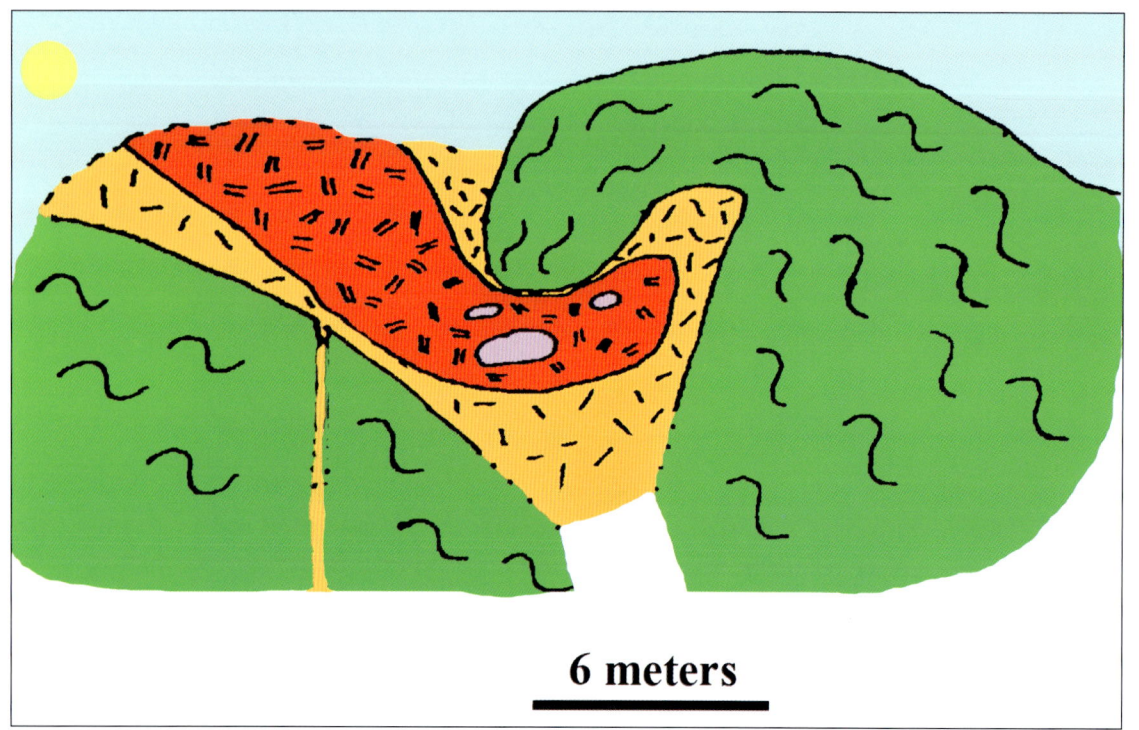

Large tourmaline pockets (pink) under a roll of a fold of schist (green). Orange = Normal pegmatite. Red = Replacement cleavelandite. Redrawn from King (1980). Pale Purple = Pockets.

6 meters

Do Minerals Point the Way to the Pockets?

Minerals and the structures inside of a pegmatite may point the way toward the interior of the pegmatite. The tourmalines shown in the Wall Zone section show wider growth toward the pegmatite interior. Minerals expand and grow into the uncrystallized fluid and merely point the way toward a portion of the pegmatite where fluid was prevalent. Closer toward a pocket, tourmalines may grow with a fan-like development toward open volumes, but they may also be encased in minerals and any cavity opening may be preserved in some other direction. Rarely, tourmalines develop pendants showing growth spurts. The best predictors of the structure of the pegmatite include observing the zonal development in the pegmatite. The direction toward pockets is not always obvious and many pegmatite miners have blasted into a pocket zone unexpectedly. In fact, an abrupt transition from "normal" pegmatite to replacement mineralization satisfies the definition of late mineralization of pocket "zones" cutting earlier formed minerals.

Jerry Cinnamon examining overlapping cleavelandite units above a large tourmaline pocket, Newry, Maine.

Pendant growth on schorl crystal. 10 x 15 cm. Emmons Pegmatite, Maine.

Fan of red elbaite in gray purple lepidolite. 12 x 17 cm. Black Mountain Pegmatite, Maine.

Minerals on the Edge
of the Pocket Zone

Pocket zones consist of different mineral assemblages than the older outer zone minerals, but the distinction is that Pocket Zones cut across previously existing minerals. Crystal cavities are occasionally found in Intermediate Zones but the walls of the cavities are the same as the minerals in that zone. The replacement pockets are formed after the remaining pegmatite fluid has become enriched in water and rare elements.

Cleavelandite lamellae in overlapping texture. 4.5 x 8 cm. Fillow Pegmatite, Branchville, Connecticut.

Cross-section of cleavelandite blades enclosed in smoky quartz. 15 x 20 cm. Crooker Gem Pegmatite, Maine.

Granite pegmatites are rocks principally composed of silicates: quartz, feldspar, mica, etc. with a small percentage of non-silicate minerals. In fact, nearly three-fourths of the granite pegmatite-forming fluid is silica. There is water among the original chemical components, but water's starting concentration is frequently less than 1% of the fluid. During the process of crystallization, water may be enriched to about 11% concentration in the fluid, but this is a maximum amount for the normal pegmatite-forming fluid. If water concentration continues, the water begins to form a separate fluid of its own probably in a sudden rather than gradual event. Many writers have used the oil and water analogy to discuss how two fluids can be in contact and be separate. The separate water-rich fluid does contain some silica and many of the rare elements are more soluble in the hot water-rich fluid than the co-existing silica-rich fluid. The principal reason replacement occurs is that the action of the water-rich fluid is very different than the silica-rich fluid. The water-rich fluid may preferentially dissolve solid rock and the rare elements in the water-rich fluid may begin to crystallize an entirely new sequence of minerals. Few detailed descriptions of what actually happens have been made, because the great number of reactions involved make it difficult to keep track of what is happening and when various reactions occur.

The mineral collector and miner should be thoroughly familiar with the following minerals and it is recommended that a reference collection is formed with numerous examples of the following species from a wide variety of locations. Some features will be briefly noted of important minerals, but text and photographs are no substitute for handling these minerals and knowing the proper way to identify them. Because of the limited space provided here, the important pocket lining minerals are not described in detail and every miner and mineral collector should have a library of mineral books as well as good magnifiers, hand lenses, and a microscope.

Feldspars

Cleavelandite: One of the very common minerals found in a pocket zone is cleavelandite. Cleavelandite was named as a platy variety of albite in 1823 in honor of Parker Cleaveland, a mineralogist who had written a very popular textbook of mineralogy that had just been published in a second edition. The original cleavelandite had been found at North America's first rubellite occurrence at Clark Ledge, Chesterfield, Massachusetts, in about 1811. Cleavelandite owes its characteristic appearance due to its rapid growth in two directions forming intergrown plates. Cleavelandite may be found in a pegmatite that didn't form pockets, but pockets in replacement zones are rarely without cleavelandite. Normal albite also may be present in the pocket zone and the typical nearly right angle cleavages may be used to identify it. A second feldspar called microcline may be present, but it does not show the platy characteristic of cleavelandite. Albite cleavages may show finely spaced striations on cleavage surfaces while microcline cleavages will not.

Cleavelandite from a gem pocket interior. 7 x 8 cm. Zé Pinto prospect, Brazil.

Cleavelandite usually forms white to occasionally pale blue warped and undulating, somewhat parallel interlocking plates. Originally cleavelandite was only named for this form of overlapping texture, but when cleavelandite crystallizes into a gem pocket, the lamellae may develop crystal faces. Rounded clusters of cleavelandite "crystals" are sometimes referred to having "cauliflower" shape.

Cleavelandite "cauliflower" from a gem pocket. 8 x 8 cm. Pederneira Pegmatite, Brazil.

Micas

Muscovite: Muscovite may be particularly abundant in a pocket zone and large "books" may signal the proximity of a gem pocket in the interior zones of a pegmatite. Micas may be found with a many intergrown minerals revealing the concentration of rare elements in the rock. Tourmaline, columbite, etc. may be present. In a few pegmatite districts, green or blue tourmaline may be found cutting across the sheets of mica, but most of the non-black tourmaline encountered will be found at the edges of the mica.

Lepidolite: There is a variety of micas of interest to pegmatite miners, particularly lithium-bearing micas, but the ones of gem importance are lepidolite and cookeite, respectively. Lepidolite is ideally colorless when pure, but it is so seldom pure that colorless lepidolite is rare. Most commonly lepidolite is pinkish to reddish purple, but it may also be bluish purple to grayish purple, although it may be yellow and indistinguishable from yellow fine-grained or minutely crystallized muscovite. The various pink shades of lepidolite are attributed to varying amounts of manganese. Gray lepidolite occurs when there is still some iron available to substitute in the mineral. When masses of lepidolite are encountered, folklore holds that a gem tourmaline pocket may be only centimeters away. There is some validity to the belief and oftentimes it is true, but many lepidolite masses are not next to any pocket or the pocket may not contain tourmaline. Lepidolite may merely seem to stain the rock or may form masses up to a meter across. In rare occurrences, such as at the Stewart Pegmatite in Pala, San Diego County, California, lepidolite has been found by the ton, frequently hosting bright pink tourmaline. The original (type) locality for lepidolite is the Rožná Pegmatite, Czech Republic and it was also mined for specimen and ornamental lepidolite with crystals of pink tourmaline, but gem pockets were essentially non-existent there.

Lepidolite that is part of the replacement process is generally fine-grained. It may show small flakes or be fine enough in texture to be fashioned into lapidary objects and carvings. Sheet lepidolite and large flake lepidolite certainly grew relatively slowly. Lepidolite may be found in small crystals and may be present around gem crystals that formed in the pocket interior, but genuinely well-formed lepidolite crystals are rare. During the final stages of pegmatite crystallization, when lithium is much enriched, lepidolite may be formed as an overgrowth on muscovite. These overgrowths almost always occur in a replacement zone near or in a crystal pocket, although there may be false signs, as with any indicator mineral.

Lepidolite with fine-grained texture. 7 x 9 cm. Mount Mica Pegmatite, Maine.

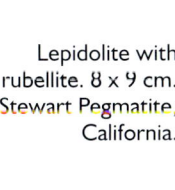

Lepidolite with rubellite. 8 x 9 cm. Stewart Pegmatite, California.

Rose Colored Muscovite: Because lepidolite is almost universally a shade of rose purple or blue purple, any fine-grained rose colored micas may be misidentified as this mineral. Rose purple muscovite probably is much more prevalent than is known and there are several other micas which may be pale purple. Large flake muscovites with a rose coloration are simply called rose muscovite, while very fine-grained rose purple muscovite may be called damourite. Damourite is commonly a replacement of spodumene or colored tourmaline and may be so fine grained that it looks like a primary mineral.

Lepidolite crystals with albite. 15 x 15 cm. Dunton Pegmatite, Maine.

Curved lilac lepidolite. 7 x 7 cm. Dobra Voda, Czech Republic.

Rose muscovite. 7 x 10 cm. Harding Pegmatite, New Mexico.

Purple damourite replacing spodumene. 8 x 8 cm. Tamminen Pegmatite, Maine.

is not usually very abundant in pegmatite-forming fluids, it may be consumed by early minerals and eventually only the white muscovite may form from the time that iron is too low in concentration to form annite. There are of course exceptions and some black mica may form during different time periods of crystallization. A "book" of mica may have a black mica core with a white mica overgrowth or vice versa. Rarely, a muscovite may have a brown to brownish yellow color zone where the muscovite has a significant iron component. Each color change or overgrowth is indicative of some change in the crystallization of the fluid. Muscovite may also form a fringe growth of small scales on a large muscovite crystal. The only implication is that there was a hiatus in muscovite growth conditions followed by a period of rapid growth of small crystals having nucleated along a muscovite crystal face.

Color-zoned Micas: Minerals may visibly record changes relating to their growth history. Changes in growth conditions may be seen in general change in crystals size and small crystals of a mineral may grow on large crystals of the same species. Crystals forming rough and cloudy crystals may greatly improve in transparency and perfection of shape when growth conditions changed. Minerals which easily accept a variety of impurity elements may become colored or lose their color depending on the availability of elements in the nurturing fluid they were growing in. Tourmaline is a good example of color changes in a mineral during growth. Micas may show color changes within themselves or a different mica species may overgrow another.

Black mica: Iron-rich mica, annite (sometimes called biotite), may form very early in a granite pegmatite's evolution. Annite is the common black mica in pegmatites. Because aluminum is so abundant, it may produce the common white mica, muscovite, but there is usually no mixing the components of annite and muscovite and the two micas may form side by side. Muscovite can have a small iron component and in that case the color is no longer silvery white, but may be yellow or even greenish yellow. Because iron

Annite core crystal with a yellow overgrowth of muscovite and a slight reddish brown overgrowth of muscovite. 8 x 8 cm. Wheeler Quarry, Maine.

White fringe of muscovite on muscovite crystal. 6 x 8 cm. Emmons Pegmatite, Maine.

Reddish purple lepidolite rim on yellow muscovite. 8 x 6 cm. Ambatofinandrahana Pegmatite District, Madagascar.

Muscovite showing alternating color zones from a crystal pocket. 7 x 8 cm. Bumpus Pegmatite, Maine.

Thin purple lepidolite rim on yellow muscovite. 9 x 7 cm. Crooker Gem Pegmatite, Maine.

Cookeite: Cookeite is not a true mica, but is a lithium-rich member of the chlorite group and has a more complex chemical formula than lepidolite. The mineral was named for Josiah Cooke, a chemistry and mineralogy professor at Harvard University in the nineteenth century. (Cooke is particularly remembered for his research on atomic weights of the chemical elements and one of his students was awarded a Noble Prize for work on atomic weights.)

Cookeite has the typical crystal clustering of all chlorites. The cookeite crystals are arranged in a radial pattern of hexagonal looking plates. Because the clusters are usually only several millimeters across, rounded clusters of muscovite may be erroneously misidentified as cookeite. Cookeite is a better indicator than lepidolite of tourmaline-bearing gem pockets, although it is generally inconspicuous and occurs in relatively few worldwide locations compared to lepidolite. Cookeite has yet to be synthesized as that mineral alone, but forms along with quartz crystals. In natural assemblages, cookeite is also found formed at the same time with quartz crystals, supporting what has been observed in the laboratory.

Cookeite casts of tourmaline, 5 x 7 cm. Mount Mica Pegmatite, Maine.

Characteristic radial clusters of cookeite.
2 x 3 mm. Mount Mica Pegmatite, Maine.

Cookeite on cleavelandite. 5 x 8 cm. Jenipapo Pegmatite, Brazil.

Tourmaline

Black Tourmaline: Of course, one of the best indicators of tourmaline is tourmaline itself. A granite pegmatite fertile enough to have gem tourmaline pockets will have lots of black tourmaline in the outside zones. Black tourmaline is rarely pure black and small glints or internal reflections best seen under magnification reveal that most black-appearing tourmaline is dark green to dark blue.

The rare essential ingredient in all tourmaline is boron. If a pegmatite has little or no evidence of boron in the form of black tourmaline, it probably won't have much tourmaline if any replacement zone mineralization is encountered. There are rare exceptions, of course. By far, the most common black tourmaline known from granite pegmatites is schorl. Sometimes, portions of tourmaline crystals are deficient in sodium and iron and relatively rich in aluminum, resulting in another species called foitite. Foitite is also black, but occasionally blue black, purple black to dark purple, dark gray, etc. Foitite is commonly a fibrous overgrowth on the tips or prisms of other tourmaline species, although entire crystals of foitite are known. Dozens of foitite localities are currently recognized, but there are many thousands of schorl localities.

Black tourmaline is iron-rich and tourmaline is very efficient in incorporating iron into its crystal structure. Iron is also a very strong coloring agent, but in very small amounts, iron may color tourmaline green or blue. All that is needed is to reduce the amount of iron available and let lithium and aluminum take the place of iron and brightly colored tourmaline crystals may form, if everything else is favorable. Additionally, if a replacement zone is likely to contain gem tourmaline, there is usually a significant display of lithium mineralization. Spodumene and triphylite might be far from the pocket, with lepidolite, montebrasite, and elbaite near the pocket margins.

Elbaite and Related Gem Tourmalines: Elbaite is the most common of the brightly colored tourmalines found in granite pegmatites and is therefore the most abundant gem tourmaline. Species such as rossmanite, liddicoatite, and olenite look like elbaite, although they are very uncommon as in the comparison between schorl versus foitite. Dark green or

dark blue tourmalines may approach or reach the tourmaline species schorl in chemical composition, although they have been overenthusiastically called elbaite, because they sell better. Brightly colored tourmaline is nearly universally present in the gem pocket lining, if there is also brightly colored tourmaline inside the pocket. Unfortunately, brightly colored tourmaline may be in a crystal pocket matrix and the interior of the pocket may contain no tourmaline.

Abundant black tourmaline near the contact of the Emmons Pegmatite. A tourmaline pocket was found about three meters below this concentration of black tourmaline, although the pocket was not considered a very rich tourmaline pocket. FOV = 1.5 x 2 meters.

Abundant black tourmaline developed above a white layer of aplite near a Pocket Zone. FOV = 1.5 x 2 meters. Bennett Pegmatite, Maine.

Green elbaite tourmaline with cleavelandite; purple lepidolite and gray quartz near a small tourmaline pocket. 0.6 x 1 meter. Mount Marie Pegmatite, Maine.

Black tourmaline with green and red color changes at the end. 60 x 14 cm. Mount Marie Quarry, Maine.

Giant tourmalines showing color changes. Black tourmaline forms when iron is present. As the amount of iron decreases, blue tourmaline may appear, and with falling amounts of available iron green tourmaline may appear. Only when iron is extremely low can pink tourmaline be formed. The concentric color zones are then recorders of changing chemical concentrations in the last stages of pegmatite crystallization. Tourmaline fan in cleavelandite and lepidolite. 55 x 15 cm. Mount Mica Pegmatite, Maine.

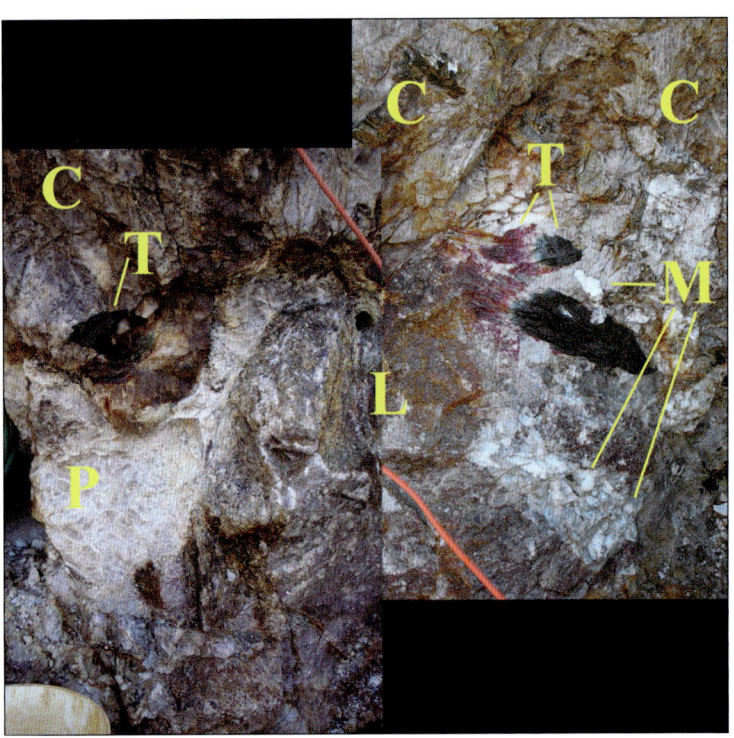

Face of a replacement zone. Top: Coarse cleavelandite "C". Center Right: White masses or nodules of montebrasite next to black tourmaline "T" with color change to green and then red extending into grayish purple lepidolite "L". Rare mineral on left, pollucite "P", demonstrate the extreme concentration of the element cesium, one of its essential components. Just above the crook of the "V" of pollucite there is a mass of green and blue tourmaline "T". The bottom point of the pollucite "V" nearly touches intermediate zone pegmatite. (Orange electrical power cord is visible in the two images.) Mount Mica Pegmatite, Maine. FOV = 3 x 3 meters.

Light and Dark Lithium Minerals

Spodumene has already been mentioned as a light colored lithium-bearing mineral. It usually forms ashen white to pale creamy yellow bladed crystals at least 0.3 meters long, but masses have exceeded 15 meters in length. An important feature includes the two very well developed, sharply angular intersecting cleavages. Spodumene may be first found in Intermediate Zones as well as later in Replacement Zones. Some miners do not want spodumene to be well developed in the Pocket Zone: "It's the kiss of death for gem tourmaline pockets" – Frank Perham, personal communication, 1988. Any lithium-bearing mineral consumes that element. If spodumene or other lithium minerals form in too great a quantity, the amount of lithium available to form gem tourmaline is greatly diminished.

Montebrasite and triphylite are also lithium-bearing minerals that are good signs, although they may also "compete" with other minerals for the available lithium in the pegmatite fluids. Montebrasite is part of a series of minerals and has hydroxyl in greater concentration than fluorine. Its fluorine-dominant relative is amblygonite, although amblygonite is rare. Montebrasite is usually bright white and may form blocky to irregular masses in the pocket zone. It may be found imbedded in cleavelandite or lepidolite. Well-developed crystals are rare and are much prized. Montebrasite has one very well-developed cleavage, while feldspar has two very well-developed cleavages. A pure piece of montebrasite will feel somewhat heavy for its size. Montebrasite has the peculiar property of being easily fusible and will bubble and melt when a small sliver is held in a flame such as from a plumber's torch and the flame around the piece may show a slight red coloration indicating lithium. Masses of montebrasite several centimeters across are considered good signs, but very large masses (e.g. 1 meter) indicate high consumption of lithium.

Montebrasite: Bright white mass with light gray quartz. Strickland pegmatite, Connecticut. FOV = 5 x 5 cm.

tourmaline are much greater. Again, lithiophilite is certainly no guarantee that any tourmaline will be found, as in the famous example of the Fillow Pegmatite in Branchville, Connecticut. The Fillow Pegmatite is the original location for lithiophilite and produced thousands of lithiophilite specimens, but it did not yield much tourmaline of any color, much less any gem tourmaline pockets. The same may be said of the pegmatites in the Custer area of South Dakota where every pegmatite seems to have at least some triphylite, but despite the district's obvious lithium mineralization, gem pockets are virtually unknown there. As with the other "indicator" minerals, many of them may be present in a particular pegmatite, but gem pockets may still prove elusive.

Montebrasite: Reverse side of specimen showing yellow-stained blocky crystal. FOV = 8 x 5 cm. Strickland pegmatite, Connecticut.

Bright brownish orange Lithiophilite originally found in large pods in cleavelandite. 6 x 7 cm. Fillow Pegmatite, Connecticut.

Montebrasite: White and creamy white masses. Large example is from Stewart Pegmatite, California. Small sample is from Bikita, Zimbabwe. FOV = 3 x 5 cm.

Triphylite and Lithiophilite: The triphylite-lithiophilite series contains essential lithium and phosphorus with either iron (triphylite) or manganese (lithiophilite). These minerals may be in the Intermediate Zone or the Pocket Zone, sometimes both, but if lithiophilite is in the pocket zone, the chances of finding red tourmaline instead of only green

Gemmy dark gray brown triphylite. 5 x 6 cm. Boa Vista, Brazil.

There may be a color difference between triphylite and lithiophilite. Iron-rich triphylite tends to be gray, greenish gray, or brownish gray, while very manganese-rich lithiophilite may be orange pink to amber orange. Intermediate compositions may be dark neutral brown to grayish coffee-brown. These phosphates are susceptible to alteration either during the crystallization of the pegmatite or due to the action of surface weathering. Triphylite may have a definite blue color because vivianite was formed as microscopic veinlets in it. Both triphylite and lithiophilite may have similar colors when they have intermediate iron to manganese ratios.

Triphylite or lithiophilite may have visibly increased porosity where late stage replacement fluids have dissolved the mineral leaving cavities with newly-formed microscopic minerals behind. The secondary minerals derived from the original phosphate mineral are highly prized by mineral collectors especially those who are interested in rare minerals and who study such specimens using a microscope.

The original formation of triphylite required lithium from the pegmatite-forming fluid. The triphylite or other lithium mineral may act as a temporary reservoir for lithium, which may be leached into late stage water-rich fluids, and the lithium then may be available for subsequent mineral formation such as montebrasite, spodumene, or even gem tourmaline. However, if only a small amount of dissolution of triphylite occurs, only a tiny quantity of lithium might have been released. What is important is the action more than the quantity. Late stage fluids that are water-rich and corrosive also form replacements and pockets. The amount of actual lithium released for re-crystallization probably has little influence on the total amount of gem tourmaline that might form, but the alteration reaction is an indicator for the miner to be aware of.

Triphylite or lithiophilite may be simply leached and oxidized without apparent dissolution. The change is a pseudomorph (false form) where the mineral retains its original form, but now has a different makeup. If the only change is oxidation of the iron or manganese along with the leaching of lithium, the resultant mineral is called heterosite, if it is iron-dominant, and purpurite, if it is manganese-dominant, although manganese is much more resistant to oxidation than iron. Interestingly, the mineral may look black

or rusty, but if the mineral is scratched or powdered, it has a dark purple coloration. Occasionally, these minerals may be unnaturally treated in various acids and the exterior of the samples then become bright purple.

Triphylite crystal (1 x 1.5 cm) with vivianite staining with quartz. G. E. Smith Pegmatite, New Hampshire.

Triphylite with minor red almandine. 7 x 7 cm. White Elephant Pegmatite, South Dakota.

Heterosite formed by oxidation and leaching of triphylite. Iron and manganese were oxidized and lithium was removed. 4 x 6 cm. Tip Top Pegmatite, South Dakota.

Purpurite derived from lithiophilite. 7 x 11 cm. Goabeb 63 Farm Pegmatite, Namibia.

Niobium and Tantalum Minerals

While the pegmatite miner and mineral collector may welcome the sign of dark minerals, it should be remembered that only a few of these dark species contain the necessary lithium to suggest the nearby presence of gem tourmaline. Dark colored minerals may be found in all of the pegmatite zones and there many to chose from in granite pegmatites, but niobium and tantalum oxides are very good species to watch for and for that reason there should be a good introduction to these species.

There are other dark minerals such as magnetite, biotite, rare-earth bearing minerals, schorl, zircon, uraninite, etc. that also need to be learned, but those species do not offer much predictive encouragement, except that schorl indicates the presence of the fundamental nutrients necessary for gem tourmaline. Knowing how to recognize minerals that may be present in replacement units in granite pegmatites will contribute to any miner's success. There are many black minerals found deeper into a pegmatite and that may be better indicators of nearby gem pocket development, although some of these minerals may also be present in an earlier zone. The most commonly encountered black and brown minerals other than schorl, some phosphates, or black stains encountered in the margins of the Pocket Zone are: cassiterite, uraninite, zircon, members of the microlite group, columbite group, tapiolite group, or wodginite group. These latter minerals are usually good indicators because they are evidence for enrichment of tin, uranium, niobium, tantalum, and sometimes manganese and evidence for the rare element enrichment the miner is hoping for. The evolution of pegmatite fluids is important, because the elements present in gem minerals such as tourmaline must be in high enough concentration to permit their formation. Nonetheless, indicators are no more than that. There have been gem pocket-producing granite pegmatites with few mineral indicators that predicted the gem pockets, but the number of examples is certainly small. Conversely, wonderful mineral predictors may be present near a crystal pocket, but the pockets may contain no large crystals whatsoever. Virtually every great gem pocket-bearing granite pegmatite in the world has some barren pockets.

Columbite-(Fe). 60 x 47 x 27 cm. Doce Valley pegmatites, Brazil. *Photo courtesy Joseph Freilich.*

Columbite-(Fe). 1.7 x 2 cm. Columbite Prospect Pegmatite, Mocksville, North Carolina.

Columbite-(Fe). FOV = 2.5 x 3 cm. Oak Hill Pegmatite, Standish, Maine (*AMNH Collection*).

Despite the uncertainty of riches, every indicator of rare element enrichment is welcome in the exploration and mining of gem-bearing granite pegmatites. Columbite Group minerals do show that the tantalum to tantalum + niobium ratio increases during the evolution of the pegmatite-forming fluid as well as the enrichment of Manganese in the Manganese + Iron ratio, but the presence of tantalite-(Mn) alone is sufficient to demonstrate that the enrichment trend is favorable. Again, indicators are no more than that. The proof in finding gem pockets only comes in their discovery.

As can be seen in the gallery photos, columbite varies in shape. Crystals may be bladed, blocky, or columnar. Low-iron columbites and tantalites may be brown to red in color, but some manganese dominant specimens may still have appreciable iron in them and thus appear black, but may have reddish brown glints. Unfortunately, color is not a good way to identify minerals. Columbite-(Fe) crystals usu-ally have a black streak and while low-iron columbite-(Mn) may have a brown streak or red-brown internal glints seen under magnification, high-iron columbite-(Mn) could easily have a black streak. This is unfortunate, as the miner wants to be in a high manganese-assemblage. Tourmalines are also sensitive to the iron-manganese ratio. High iron tourmalines are black. Low amounts of iron in tourmaline may permit very dark green or very dark blue tourmaline to form. When iron is very low, bright green or blue tourmalines may be present. Only when ferrous iron is almost absent in the pegmatite fluid are very pale shades of tourmaline likely to form. Rubellite is essentially iron-free. Of course, other nutrients such as lithium are needed to form elbaite and there are further complexities in tourmaline formation. The rare associated dark minerals within the "pocket zone" should not be ignored, especially as they have retail value. Because of the real value of dark minerals, only broken fragments should be crushed for testing.

Columbite-(Fe) with quartz and cleavelandite. 7 x 9 cm. Wentworth Pegmatite, Maine.

Columbite-(Fe) with blue tarnish. 1 x 3 cm. Guy Johnson Pegmatite, Maine.

Columbite-(Fe). 1 x 3.4 cm. Golconda Pegmatite, Brazil.

Microlite, another tantalum mineral, is a very favorable species indicator of enrichment and it has a distinctive appearance. Most commonly, they have an octahedral shape and often there are tiny modifying faces with the dodecahedron. Other faces, such as the cube, are rare. From this description, it is immediately obvious that pegmatite miners need to know crystallography. Microlite may be yellow, tan, brown, and almost black brown. Occasionally, small amounts of uranium may be present and there may be a slight brown color halo in the matrix where radioactivity has affected the minerals in contact with it. Some pegmatite miners own microscopes to examine indicator minerals, because small minerals carry important "information".

Tantalite-(Mn) with quartz and cleavelandite. 7 x 9 cm. Dassu, Pakistan. *Photo courtesy of Stuart Wilensky.*

Tantalite-(Mn) crystal. 3.3 x 3 cm. Bennett Pegmatite, Maine.

Tantalite-(Mn). 2.5 x 5 cm. Darra-i-Pech, Afghanistan. *Photo courtesy of Russ Behnke.*

Mineral Paragenesis

Paragenesis simply refers to which minerals were formed through time, although some species may be crystallizing at the same time as others. Many miners and mineral collectors think that paragenesis simply means an inventory list of the minerals that are found together; however, a specimen may easily have a variety of minerals found together, but which formed far apart in time from each other. The discussion of zones and replacement in granite pegmatite is a study in parageneses. Zones are big paragenetic features of pegmatites, but there are many subtle mineral changes that do not have such a gross scale. The distinction is that the minerals of a paragenetic stage came from a similar fluid. In practice, a mineral specimen may have representative minerals from a variety of overlapping parageneses. The matrix may have the early minerals that crystallized together and there may be several later episodes when newer minerals were formed and even in cavities there may be a definite sequence of minerals which are found together, but which did not form under precisely the same conditions and times. Minerals may be associated with each other on a specimen, but because they formed much before the other minerals, they are associated only by the accident of where they are. The reasons for introducing the term paragenesis this late in the discussion about granite pegmatite formation include concerns of putting the "cart before the horse." With a background in understanding that there are zones in granite pegmatites, the miner will understand paragenesis intuitively.

Octahedral yellow microlite, pale creamy white montebrasite (right), cleavelandite (upper), and light gray quartz. FOV = 7 x 7 cm. Dunton Pegmatite, Maine.

Tantalite-(Mn), Nuristan, Afghanistan. 11 x 4.5 cm. *Photo courtesy of Joseph Freilich.*

Dark brown microlite with green elbaite, pale lepidolite, and cleavelandite. FOV = 4 x 7 cm. Gillette Pegmatite, Connecticut.

Brown microlite in a Lithium-rich mineral assemblage. 6 x 9 cm. Purple lepidolite, light pink fine-grained muscovite, creamy yellow blades of spodumene, white cleavelandite (bottom and short white blades in lepidolite), white montebrasite (upper right edge), black and gray smoky quartz (lower edge), yellow brown microlite, and black tantalite-(Mn). Harding Pegmatite, New Mexico. Despite the abundance of favorable indicator minerals, the Harding Pegmatite is not known for having produced gem pockets.

Geologists studying pegmatites early recognized that there were changes that revealed the sequence of crystallization of the rocks. Waldemar Brøgger (1890) studied nepheline syenite pegmatites and wrote an important book length article on the sequence of crystallization in rocks exposed along the Norwegian coast. Brøgger described the succession of minerals that formed and this article greatly influenced everyone who studied pegmatites afterwards. Brøgger's ideas were not new, but he eloquently documented and described the details of mineral history. Among his contributions was the explanation that the minerals that formed at any one moment were related to each other and that the earlier or later formed minerals were from different fluids.

In 1925, a student, Kenneth Knight Landes published the results of his Ph. D. studies. He recognized that there were many changes in granite pegmatite formation including the dissolving and recrystallization of pegmatite. In succeeding years, geologists recognized the various mineral-forming (paragenetic) stages in granite pegmatites and after the great exploration of granite pegmatites during World War II, a formal classification of the mineral changes in granite pegmatites was published by Cameron, et al. (1949).

Various diagrams have been devised to illustrate paragenesis. The most common type shows only the overall features of crystallization. The diagram from Woodard (1951) shows the major minerals of the Lord Hill pegmatite and a few minor species. Such paragenetic diagrams illustrate the changing mineral composition of solid granite pegmatite as the pegmatite-forming fluid evolved. There are abrupt mineral changes shown on the diagram and these changes can be seen as distinct zones. Some of the common minerals persist throughout the history of the crystallization of the pegmatite while some crystallizing minerals appear or disappear through time. There are similar paragenetic diagrams for other kinds of mineral deposits and sulfide mineral studies have also benefited from these ideas.

The paragenetic diagram shows the presence of a mineral as a colored band. If the mineral is rare, only a line's thickness is shown. Very abundant minerals are shown by a thick band more or less corresponding with the mineral's percentage in the rock. The scale from left to right represents time with the first formed and therefore oldest minerals on the left with the youngest minerals on the right. The diagram does not show the amount or proportion of time, which passed from oldest to youngest mineral formation, but probably some of the most interesting minerals crystallized quickly over a short time interval. Woodard (1951) lumped the Wall and Core Zones together, probably as not much core was then exposed in that pegmatite. Interestingly, the diagram shows replacement and cavity formation as separate features. More of the pegmatite was exposed during mining after Woodard's study. In the late 1950s and the early 1960s, numerous crystal cavities to 3 x 4 x 3 meters were found, including smoky quartz crystals weighing over 100 kg and a well-terminated North American record topaz crystal weighing 25 kg.

Although there are over 800 minerals known from granite pegmatites, it may be astonishing to know that about half of the minerals found in granite pegmatites are phosphates. Particularly, granite pegmatites may have masses or pods to several meters in diameter of phosphate minerals such as triphylite-lithiophilite, apatite, or montebrasite and these masses may be partially replaced by later minerals derived from the recombination of the dissolved phosphate with components then present in the pegmatite-forming fluid. The study of replaced phosphate masses is so popular among mineral collectors that it should be discussed in a little detail, especially as many of the minerals are so attractive, although most of the species are microscopic.

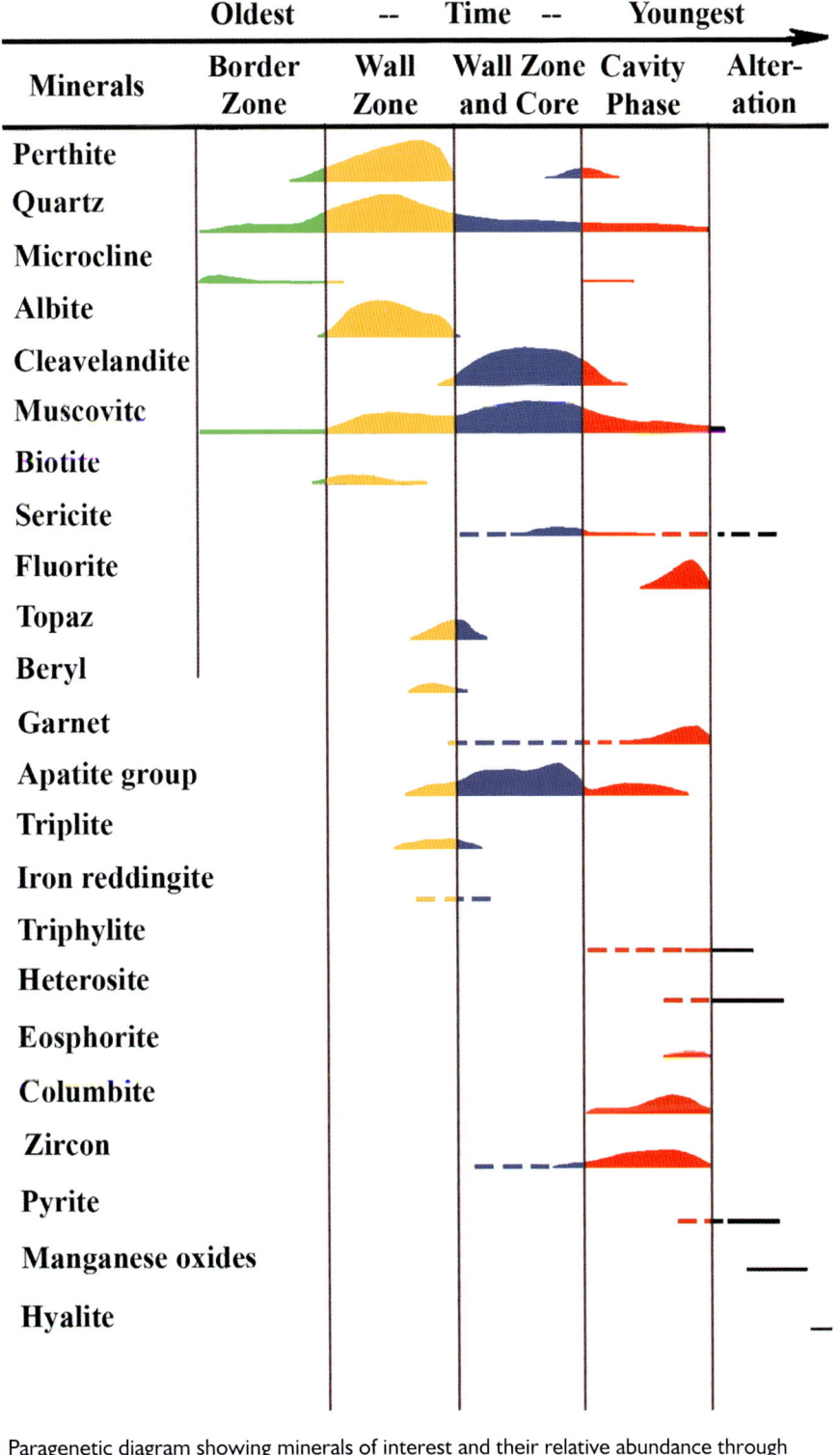

Paragenetic diagram showing minerals of interest and their relative abundance through time in the Lord Hill Pegmatite, Maine. Granite pegmatites crystallize from the outside toward the inside, so the oldest minerals are on the outside and the youngest are inside. Changes in mineral proportions and kinds of minerals that formed may be seen and these observations are the basis of recognizing and assigning zones. Redrawn from Woodard (1951).

Of the various phosphate pods, triphylite-lithiophilite is the starting (primary) material for many of the hundreds of secondary granite pegmatite phosphates and provides an excellent case study of paragenesis in a very limited part of a granite pegmatite. Dramatic mineral changes may occur across only several centimeters or even millimeters. Extreme alteration of triphylite-lithiophilite might result in most or all of the phosphate being removed from the original mass with only dark colored black and brown minerals remaining. The paragenetic diagrams of Moore (1973) are still valid and important to anyone interested in these fascinating species.

In the process of phosphate pods being dissolved and replaced, additional minerals may be deposited which had hitherto been absent in the granite pegmatite. Siderite-rhodochrosite is commonly found replacing triphylite-lithiophilite and indicates that carbonate was a late-stage component in the corrosive fluids. The siderite-rhodochrosite often occurs in masses ranging from several millimeters to over a meter across and has crystallized cavities into which secondary phosphates may form. The siderite-rhodochrosite often occurs in masses ranging from several millimeters to over a meter across and may have crystallized cavities into which secondary phosphates may form.

Phosphate pod from a pegmatite: The original material is gray triphylite with a blue tinge due to the impregnation of vivianite along cleavage fractures. The upper rind has a replacement by siderite. Just inward there is the development of a black mineral, rockbridgeite, with some of the last formed minerals in the interior, including orange laueite and tan strunzite. 9 x 12.5 cm. Palermo #1 Pegmatite, New Hampshire.

Either within cavaties of the siderite-rhodochrosite or within the triphylite-lithiophilite,there may be phosphate minerals present that are iron or manganese dominant, just as the parent triphylite or lithiophilite, respectively, were iron or manganese dominant.

Iron and manganese may fit about equally as well into the same sites in a crystal and these elements have about the same size and electronic charge. For these reasons, early-formed secondary phosphates may have about the same ratio of iron to manganese as their parent primary phosphates. This fact is important to remember when identifying an early-formed secondary phosphate species. However, through time, iron is very sensitive to oxidation, while manganese is not. Water-rich late pegmatite fluids may oxidize iron and because oxidized iron (ferric iron [Fe^{3+}]) is much smaller than unoxidized iron (ferrous iron [Fe^{2+}]) and has a new electronic charge, it will not fit into a crystal site requiring the larger unoxidized iron. Because manganese is resistant to oxidation, it may become dominant in the secondary phosphates forming. Later iron-bearing phosphates may have separate crystal sites where ferrous iron is required in one and where ferric iron is required in another. The black mineral rockbridgeite is a common example of a mineral with both ferric and ferrous iron. Eventually in the process of phosphate pod replacement, iron and manganese ratios in a mineral may be dramatically different from their initial starting ratios in their primary starting materials.

Very late phosphates may have almost no ferrous iron available and spectacular minerals such as cacoxenite or strengite may form. Alternatively, the last phosphate minerals to form may be very manganese rich, such as huréaulite, or contain only unoxidized manganese plus only oxidized iron, such as the beautiful species laueite. (Manganese may stay unoxidized until the time when clays and stains are forming in a pegmatite.)

Within a microscopic area there may be minerals that 1) contain ferrous iron, followed by 2) younger phosphate minerals containing both ferrous and ferric iron, and with 3) the youngest phosphates containing only ferric iron, perhaps with unoxidized manganese. These are three distinct episodes of phosphate mineral formation based on what was happening to any iron present in solution and each represents a change in the localized character of the pegmatite-forming fluid. The discussion of the formation of secondary phosphates is much simplified and quite often there are very late reversals in the trend from unoxidized through completely oxidized iron.

Another example of phosphate mineral paragenesis is shown by a "color-coded" specimen from a phosphate vein, which unfortunately is not from a granite pegmatite. Large masses of bright green variscite were fractured and late stage fluids permeated the fractures and subsequent mineral formation formed differently colored mineral layers or shells around the earlier formed minerals. Similarly some phosphate minerals in granite pegmatites experience the same sort of replacement, but are not usually as obvious as the variscite example.

Triphylite may be dissolved and its components reacted with other elements in the pegmatite-forming fluid to crystallize a wide variety of mineral species. These four are remarkable examples of what may be found when a collector sets out to discover these minute treasures.

Cacoxenite. 4 x 6 mm. Hagendorf Süd Pegmatite, Germany. *Photo courtesy of Stephan Wolfsried.*

Phosphate pod, not from a pegmatite, that has excellent color contrast to show the sequence of mineralization. Because the pod is replaced inwardly, the core represents the oldest material and the replacement minerals are generally newer toward the outside. The green variscite is surrounded by yellow crandallite. There is a thin white layer of apatite surrounding the crandallite and gray wardite is the youngest of the phosphates. The exterior of the pod is surrounded by a mixture of minerals showing iron brown staining. 15 x 20 cm. Little Green Monster Mine, Utah.

Phosphophyllite. 12 x 12 mm. Hagendorf Süd Pegmatite, Germany. *Photo courtesy of Stephan Wolfsried.*

Hureaulite. 1 x 1 cm. Hagendorf Süd Pegmatite, Germany.
Photograph courtesy of Stephan Wolfsried.

Strengite. 1 x 1 cm. Hagendorf Süd Pegmatite, Germany.
Photograph courtesy of Stephan Wolfsried.

What Does an Unopened Pocket Look Like?

Stories about "unexpected" crystal pocket encounters are legendary. Even miners who carefully pay attention to the zonation and mineral signs can be fooled. Gem pockets may be very close to primary zones and a dynamite blast may cut into a pocket or even scatter its contents into the woods. It is not uncommon that a miner drills directly into a pocket. If a gem miner has never seen an unopened pocket, it may go undiscovered until a poacher finds it after the mining day or it gets exploded to bits by the next blast. Experienced miners reduce the quantity of explosives used as they get deeper into the pegmatite. Fortunately, there are additional warning signs, including increased porosity if even a tiny bit of the replacement minerals are exposed. Even the light color of drilling dust is important.

Within the Pocket Zone, there is frequently a quartz veining or porous texture permeating the rock. Some pegmatite miners call this a "silica phase," where they observe this association next to the pocket walls. The quartz crystals may be of any size, but the truly giant quartz crystals are found inside the pocket chamber.

One of the rarest of sights at a granite pegmatite prospect is an unopened pocket. The discovery of gem minerals excites the entire mining crew and rarely is there a photograph taken of the first exposure of a pocket. There are many pictures available of emptied gem pockets and these photographs have frequently given the impression that the wonderful specimens actually occurred in open caverns. While "cookie jar" pockets are legendary and do occur, most gem pockets are filled with "pocket sand" or "pocket mud" composed of variously fine or coarse-grained particles of albite, quartz, muscovite, etc. The illustration shows an unopened gem pocket that was filled with sand and several dozen indicolite crystals. The indicolite crystals varied from 5 mm to almost 4 cm and the largest crystal cut the North American Record for an indicolite gem (35.17 carats).

Pockets may be filled only with clay and, even within a pegmatite district, the texture of pocket filling material may vary. A pocket's filling may also be stained red, brown, or black with late mineralization. In San Diego County, California, the pocket fillings sometimes may be unstained, but are frequently stained by red brown iron minerals such as hematite or goethite. In New England, pockets may barely have a thin film of rusty coating at most. The huge "Tuesday" tourmaline pocket at Newry, Maine, with nearly pristine crystals that barely needed to be rinsed with water to remove dust was next to another huge pocket that was thoroughly stained by a blackish brown sooty manganese mineral called todorokite. Pocket clays consist of a variety of minerals ranging from ground feldspar and quartz, as well as more "normal" clay minerals such as muscovite, montmorillonite, tosudite, or kaolinite. Clay is simply a textural term for fine-grained minerals that would remain suspended for a long time in water if the mixture were shaken.

Pocket (unopened) showing spillage of contents and two dark blue indicolite crystals. Bright reflection in the black-stained pocket sand is the termination of another indicolite. Pocket cross-section = 17 x 15 x 20 cm. Crooker Gem Pegmatite, 2007.

Bob Daigle digging in a quartz crystal pocket that was somewhat cemented by quartz at Crooker Gem Pegmatite, 1992.

Porous Quartz rich vein with quartz crystals next to cleavelandite, light green elbaite, and minor light purple lepidolite. FOV = 35 x 45 cm. Mount Marie Pegmatite, Maine.

When crystals arrive in the marketplace, they have almost always gone though some process of cleaning other than just washing. Much of the time, the specimens have been put in acids to dissolve simple stains or put in special solutions that involve complex chemical reactions necessary to remove difficult to remove stains. Some crystals may have had micas or other minerals that were deposited on the crystal surface and a complicated succession of procedures may have been used to remove unwanted minerals. Often crystals are re-attached to their former growth spots.

There are also specimens that have been fraudulently created where the crystal and matrix rock were not even near each other when they were originally formed. Repaired specimens are still valuable, but fraudulently assembled specimens may not be worth the sum of their parts.

Miner cleaning out pocket sand in a tourmaline pocket.

Cookie jar pocket. 15 x 20 x 25 cm.
Bennett Quarry, Maine.

Miner cleaning pocket sand out of a tourmaline pocket. Note that the pocket is being worked with hand implements and the pocket opening has not been expanded using power tools. Cramped conditions are very common when working crystals pockets as any misdirected energy can fracture the valuable minerals.

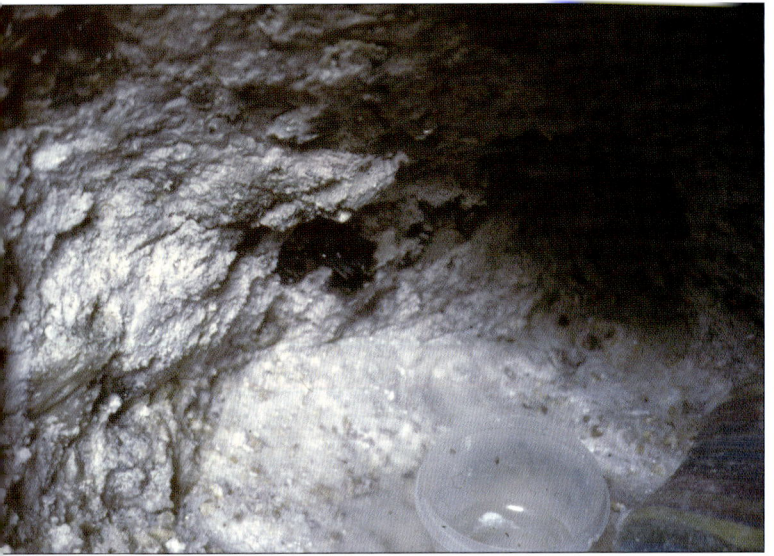

Pocket that was filled with white clay
[Lithian tosudite]. 30 x 20 x 25 cm.
Bennett Quarry, Maine.

The Pocket Lining

The quest for gem pockets has had a long lead in. You first have to recognize that a rock is a granite pegmatite. The minerals of the granite pegmatite have a visible distribution pattern with many of the rare minerals offering the encouragement that a gem pocket(s) may be near. The mineral changes that mirror the evolution of the pegmatite's parent fluid are usually abrupt and when replacement minerals are found, the miner knows that the culmination of crystallization has resulted in a prize: a gem pocket "zone." For many pegmatites, the replacement minerals are as good as the prize gets and there are no pockets with crystals or gem minerals. In a few pegmatites, the replacement minerals form a compact assemblage with some porous regions. The open space may have relatively small cavities called vugs that are only a few centimeters or millimeters across. Pockets usually differ from vugs in that there may be a systematic lining of minerals surrounding a larger cavity. There is the opposite extreme where a large cavity tens of centimeters or more across has no unique pocket lining. All too often, pockets are only filled with sandy grains.

One of the largest crystal pockets ever found in a USA granite pegmatite was at the Bennett Pegmatite, Buckfield, Maine, that had maximum dimensions of 7.6 x 2.7 x 2.4 meters (25 x 9 x 8 feet). Folklore has exaggerated that this pocket eventually reached upwards of 12 meters, but no contemporary measurements are available. This gem pocket was encountered in 1924 and took several years for the miners to partially remove its contents. The feldspar miners at the Bennett Pegmatite avoided the pocket as much as possible because the pocket and immediately adjacent rock contained few if any valuable minerals and it cost money to dynamite the pocket walls, dig out the pocket sand and occasional crystals, and move them to the dump. The pocket was partially cleaned out by Harold Perham in 1927 when he became the new leaseholder (Stan Perham, personal communication, 1968). Harold's brother, Stan, was a mineral dealer and he could sell minerals specimens while the feldspar company was only interested in a continuous supply of feldspar. In fact, persistent pockets have been detrimental to feldspar mining and "well-mineralized" portions of a pegmatite may be avoided by miners because the mineral

they were seeking was not very pure there. Later, miners Ernest Schlichter and Robert Eastman further removed remnants of this pocket in the 1970s (Ernest Schlichter, personal communication, 1990) and the final vestiges of the 1924 pocket were mined in the mid-1990s by Ronald and Dennis Holden. The only known photograph of the 1924 Bennett Pegmatite's pocket interior shows a wall of densely compact feldspar and quartz sand near the "entrance" of the pocket. A shovel in the center was probably 1.3 meters long. The end of a quartz crystal is exposed in the pocket sand and the complete crystal is now a grave marker. The pocket interior photograph shows pocket walls without a crystal lining. In fact, the texture of the walls suggests that this portion of the pocket cuts across an intermediate zone without replacing the walls by cleavelandite or other typical replacement minerals.

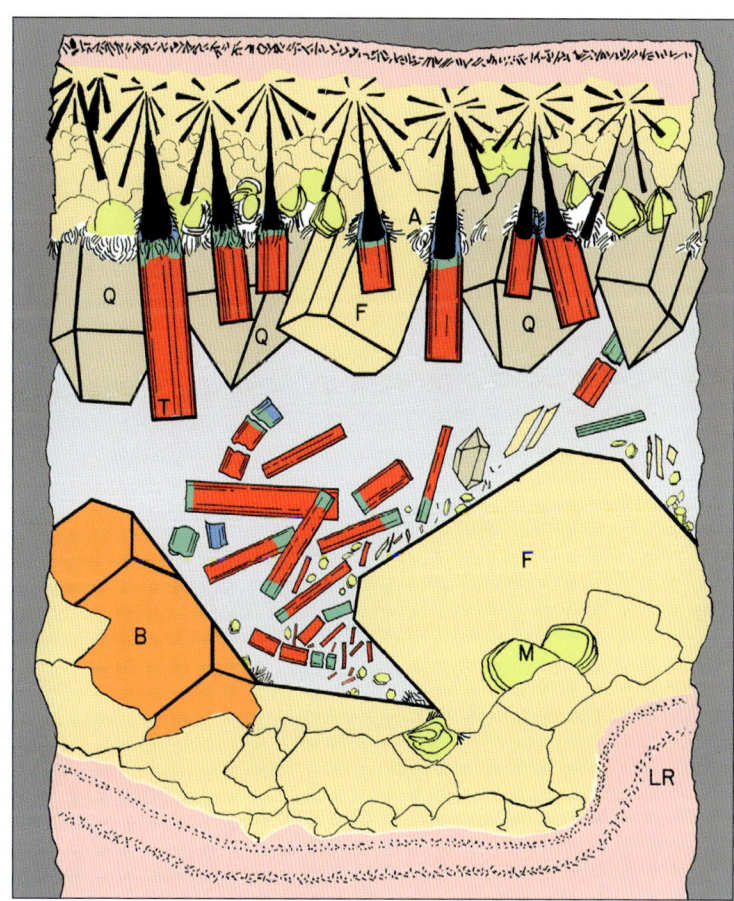

Gem pocket diagram showing idealized cross-section of the Himalaya Pegmatite, California. The walls of the pocket are well crystallized with wonderful treasures. The pocket is filled with a sandy or clayey filling and loose crystals may be found in the pocket interior (Adapted from Sinkankas, 1970). Pink margins of the diagram are aplite (LR = Line Rock).

Dennis Holden using a drill to loosen a crystallized pocket-lining specimen containing milky quartz crystals and a huge morganite crystal. This gem beryl crystal was named the "Rose of Maine." Note the large quartz crystal cluster in the back. Bennett Pegmatite, Maine. *Photograph courtesy of Wayne Flanders.*

ing, but more often than not, miners use wedges, drills, or jackhammers to further expose the cavity. In difficult cases, another small explosive charge must be used to "loosen" the rock, but such charges are very small and are designed only to crack rock enough so that hand tools can be used. When pockets are expected to have fine crystals lining the pockets, extreme care must be used otherwise very beautiful matrix specimens may be destroyed. Many pegmatites do not have desirable pocket linings and the best specimens are enclosed in a fine sand or clay. Obviously, a shovel can not be used to penetrate the sand as one scrape of bump by the shovel could destroy a fine specimen. Pockets need to be "unloaded" with the greatest of hand care. Oftentimes, pocket sand can be separated from its enclosed crystals by simple rinsing with water on a screen.

Interior of Bennett Pegmatite's huge pocket (Note root end of a giant quartz crystal), c. 1925. From Landes (1933). (Note root end of a giant quartz crystal in center right. c. 1924.)

Pockets are very irregular and the chance discovery of a gem pocket also includes the chance of what part of the pocket chamber wall is first found. Sometimes the cavity is approached from the extreme bottom with the cavity increasing in size as the miners awkwardly dig upwards, while occasionally pockets are found on their tapered ends, sides, or top depending on where the most recent explosive blast has exposed the pocket. Experienced miners realize that they must enlarge the size of the pocket opening as they need to easily remove crystals without damaging them in the process of removing them through a small rocky opening. Sometimes hand tools are sufficient to enlarge the open-

Pocket Clay and Pocket Sand

Large quartz crystal that was projecting in the 1925 pocket.

The origins of pocket sand and clay will be mentioned only in passing. Some of the pocket filling is the result of turbulent churning action of the late fluids in the granite pegmatite. The process of crystallization is not always a quiet process and there are many factors which can account for energetic movement of the fluids. The most obvious way to have violent action in a gem pocket is that the pegmatite has a small fracture. Sometimes outside movement during the formation of a granite pegmatite causes a crack in the rock to form, while at other times the inside "steam pressure" escapes through a tiny fracture because the pegmatite is contracting as it cools. Either cause will instantly let the residual water-rich fluids boil just as taking the lid off a pressure cooker would let the contents of the pot vaporize.

Emptied crystal pocket. FOV = 2 x 3 meters. Note light from a second entrance. Emmons Pegmatite, Maine.

Tourmaline pocket filled with tourmaline, pocket sand, and pocket clay, 0.7 x 1 meter. Himalaya Pegmatite, California. *Photo courtesy of Bill Larson.*

The effervescence of the boiling fluid is what caused the crystal contents of a gem pocket to break, crack, or chip and is often the reason why originally perfect crystals are damaged. After the granite pegmatite cools further, the inside pressure completely dissipates. The remaining forces acting on a pegmatite are usually the deep penetration of simple weathering. Of course, the pocket contents then quietly exist until the are exposed and there are no longer trapped pressurized fluids. When gem pockets are opened by miners, about the only fluid remaining is due to groundwater seepage into a pocket and it is not rare for large pockets to be partially filled with water.

The sand filling of a pocket is not the result of direct crystallization, otherwise the sand would be commonly cemented together by regrowth on the tiny fragments. The churning frequently resulted in ripping of crystals off from a pocket wall, the smashing together and pulverization of mineral fragments, and the thorough redistribution of pieces throughout the pocket. Many originally pristine gem crystals were destroyed millions of years before any miner got to see them and pieces of crystals may be found far from each other that fit together perfectly, showing they were once one crystal. If a pocket did not rupture, the crystallizing fluids might continue to act and tourmaline might be chemically converted into mica, etc.

The chunks and pieces found in pockets may also include minerals formed in the earliest stages of the pegmatite's evolution. Wall Zone pegmatite and graphic granite are commonly found in pockets, far away from where they "should" be. Occasionally, pieces of the host country rock are found in a gem pocket. It may be supposed that a piece of the country rock was caught up during the emplacement of the pegmatite and it was near the replacement zone and survived, but more often those pieces of rock, found where they shouldn't be, are passed over without adequate explanation.

Tourmaline and Quartz Crystal pocket with pocket filling removed and showing crystals on the pocket walls. 0.5 x 1 meter. Himalaya Pegmatite, California. *Photo courtesy of Bill Larson.*

True clay minerals, on the other hand, are formed very late in the pocket-forming process. Minerals in the replacement zone may be further replaced from the action of any pocket fluids still active. The clay forms such tiny particles that they do not frequently form a cement of the pocket sand. Occasionally, clays are the only infilling in a pocket or there may be tourmaline crystals seemingly floating in a compact clayey matrix. Clays by definition are extremely small particles and may be such minerals as muscovite, montmorillonite, nontronite, kaolinite, tosudite, halloysite, etc. and are usually white, off-white, pale yellow, but may be heavily stained red, brown, or black. Some species, such as nontronite have a light olive green to greenish brown color that is related to the fundamental chemistry of the mineral, while its close relative montmorillonite, that is usually white to tan, may have a pink color due to an impurity of manganese. Kaolinite is rare in some gem pockets, but because kaolinite forms huge sedimentary deposits, some collectors have erroneously believed that kaolinite is also common in granite pegmatites. When clays are subjected to specific tests using x-rays, the results may be surprising.

Pocket clays may be intermixed with fine-grained red to brown or black staining minerals such as hematite (red, red brown, to brownish black), goethite (beige, golden brown, brownish black), todorokite (brownish black), as well as a host of manganese-rich black minerals, possibly including cryptomelane, nsutite, birnessite, romanèchite, lithiophorite, etc. Historically, pyrolusite was thought to be common in granite pegmatites. Potter and Rossman (1978) investigated dendrites and other formations, both from granite pegmatite and other sources and they were unable to find any pyrolusite from any granite pegmatite and their findings are still true at this writing. When there are very deeply stained pocket fillings, the pockets may be essentially cemented and may be very difficult to work. Staining is very common in regions where weathering agents including ground water have seeped into pockets. Southern California; Minas Gerais, Brazil; and Madagascar are a few of the pegmatite areas where late mineral solutions have permeated crystal pockets. Great care is needed in chemically cleaning mineral stains as chemical reactions occur that may create a more objectionable and harder to remove stain.

Fine-grained muscovite "clay" coating on cavity in white beryl. 4 x 7 cm. Lord Hill Pegmatite, Maine.

Pink montmorillonite on quartz impregnated albite. FOV = 6 x 6 cm. Tamminen Pegmatite, Maine.

Bright white Lithian tosudite intermixed with pollucite. 5 x 5 cm. Crooker Pegmatite, Maine.

Quite often, fine-grained minerals grew on the exterior of well-formed pocket crystals and much effort has been directed in removing clay coating from beautiful gem crystals. Unfortunately, coatings and stains also add some verification of a mineral specimen's origins and there is a debate over the merits of cleaning a specimen. One person may want a beautiful, unstained specimen, while another may be interested in having a collection from favorite locations and the stains may be the subtle clue that validates the locality label. Most commonly, the fine-grained mineral that grew on tourmaline is muscovite, although montmorillonite, lepidolite, kaolinite, cookeite, and other clays may coat a gem crystal.

Pocket sand is easy to wash off as it doesn't have the tight clinging properties of clay. When the pocket contents are not heavily stained or coated by red, brown, or black minerals, the joy of gem pockets is at its finest. Careful digging may yield treasures in every handful or await a last moment to be revealed.

Kaolinite in quartz. 7 x 10 cm. Bumpus pegmatite, Maine.

Fine-grained creamy white boromuscovite "clay" coating cleavelandite. 3 x 4 cm. Himalaya Pegmatite, California.

Creamy white kaolinite with remnant microcline intermixed. 6 x 6 cm. Spruce Pine, North Carolina.

John Barlow at the Himalaya Pegmatite, California, examining unsorted tourmaline gem pocket contents from the current day's mining. *Photo courtesy of Bill Larson.*

Unsorted gem pocket contents. Note heavy red staining of pocket sand. FOV = 0.3 x 0.5 meters. Himalaya Pegmatite, California. *Photo courtesy of Bill Larson.*

"Mother and Child Reunion" showing cleaned tourmaline, microcline, and quartz crystals recovered from the Himalaya Pegmatite. Note how close the gem pocket shown is to the pegmatite contact and the black tourmaline fans. *Photo courtesy of Bill Larson.*

Chapter Four
Pocket Crystal Treasures

Spodumene. 7 cm tall. Mawi, Afghanistan.
Photo courtesy of Russ Behnke.

Elbaite. 7 cm tall. Malkhan District, Russia. *Photo courtesy of Russ Behnke.*

"Aqua Crown" Aquamarine. 43 x 34 cm. Shigar Valley, Pakistan. *Gene Meiran Collection.*

Topaz. 14 x 14 cm. Mimoso do Sul, Brazil.

Spessartine. 8 x 20 cm. Gilgit, Pakistan.

Chrysoberyl. 7 x 7 cm. Collintina, Brazil. *Photo courtesy of Rock Currier.*

Beryl. 12 x 5 cm. Barra de Salinas, Brazil.

Alice-blue elbaite. Rare indicolite color named for Alice Roosevelt's ball gown in 1910. 3 x 0.6 cm. Havey Pegmatite, Poland, Maine. *Photo courtesy of Russ Behnke.*

Topaz. FOV = 8 x 10 cm. Ghundao Hill, Afghanistan.

Spodumene. FOV = 8 x 5 cm. Darri-i-Pech, Afghanistan.

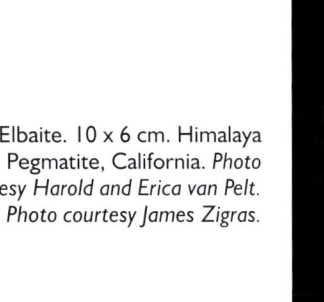

Elbaite. 10 x 6 cm. Himalaya Pegmatite, California. *Photo courtesy Harold and Erica van Pelt. Photo courtesy James Zigras.*

Rose Quartz crystals. 12 x 8 cm. Pitorra Pegmatite, Brazil.

Smoky Quartz gem. 27 mm wide (49.8 carats). Minas Gerais, Brazil. *Bill Damron Collection.*

Spessartine. 8 x 10 mm (5.94 carats). Oyo, Nigeria. *Bill Damron Collection.*

Amazonite. 20 x 20 cm. Kosnar Claim, Colorado. *Photo courtesy of Brian Kosnar.*

Purple fluorapatite. 7.5 x 5 cm. Nuristan, Afghanistan. *Photo courtesy of Russ Behnke.*

Polychrome elbaite. 7.5 cm tall. Ferruginha Pegmatite, Brazil. *Photo courtesy of Russ Behnke. Dan Record Collection.*

Rare combination of four displayable species on one piece – Topaz, Quartz, and Aquamarine on cleavelandite. 11.5 cm wide. Gilgit, Pakistan. *Photo courtesy of Russ Behnke.*

Elbaite. 14 cm tall. Mawi, Afghanistan. *Photo courtesy of Russ Behnke.*

Summary

Fred Davis scouting a pegmatite.

Miners reading the freshly cleaned rock face.

Getting equipment into a locality.

Drilling holes for explosives.

Inserting dynamite.

Moving rock.

Woody Thompson inspecting blast debris.

Remarkable purple to pink Kunzite crystal with sharply developed rose pink morganite beryl on white cleavelandite blades and light creamy smoky quartz. 20 cm wide. Mawi, Afghanistan. *Photo courtesy of Stuart Wilensky.*

References

Bastin, Edson S., 1911, Geology of the Pegmatites and Associated Rocks of Maine including Feldspar, Quartz, Mica, and Gem Deposits, U. S. Geological Survey Bulletin 420, pp. 81.

Brøgger, Waldemar Christopher, 1890, *Die Mineralien der Syenitpegmatitgänge der südnorwegischen Augit- und Nephelinsyenite*, Zeitschrift für Kristallographie und Mineralogie, v. 16, p. 1-663.

Cameron, Eugene N., Jahns, Richard H., McNair, A. H., and Page, Lincoln. R., 1949, Internal Structure of Granitic Pegmatites, Monograph #2, Economic Geology, pp. 115.

Crosby, William Otis and Fuller, Myron Leslie, 1897, *Origin of Pegmatite*, American Geologist, v. 19, p. 147-180.

Jahns, Richard H., 1953, *The Genesis of Pegmatites. I. Occurrence and Origin of Giant crystals*, American Mineralogist, v. 38, p. 563-598.

Jahns, Richard H., 1955, *The Study of Pegmatites*, Economic Geology, v. 50, p. 1026-1130.

Jahns, Richard H., Griffitts, Wallace R., and Heinrich, Eberhardt William, 1952, Mica Deposits of the Southeastern Piedmont, Part 1: General Features, U. S. Geological Survey Professional Paper 248, p. 1-99.

King, Vandall T., 1980, Distribution of Alkali and Alkaline Earth Elements in a Newry, Maine Pegmatite, M.A. Thesis, State University of New York at Buffalo, privately published, Rochester, NY, pp. 131.

Kretz, R., 1968, Study of Pegmatite Bodies and Enclosing Rocks, Yellowknife-Beaulieu Region, District of Mackenzie, Bulletin 159, Geological Survey of Canada, pp. 112 + 4 map plates.

Landes, Kenneth K., 1925, *The Paragenesis of the Granite Pegmatites of Central Maine*, American Mineralogist, v. 10, p. 355-411.

London, David, 2008, Pegmatites, Special Publication 10, Mineralogical Association of Canada, Québec City, Québec, Canada, pp. 347 + CD.

Moore, Paul B., 1973, *Pegmatite Phosphates: Descriptive Mineralogy and Crystal Chemistry*, Mineralogical Record, v. 4, p. 103-130.

Potter, R. M. and Rossman, G. R, 1978, *Mineralogy of Manganese Dendrites, and Coatings*, American Mineralogist, v. 64, p. 1219-1226.

Rickwood, P. C., 1981, *The Largest Crystals*, American Mineralogist, v. 66, p. 885-907.

Woodard, Henry H., 1951, *The Geology and Paragenesis of the Lord Hill Pegmatite, Stoneham, Maine*, American Mineralogist, v.36, p. 869-883.

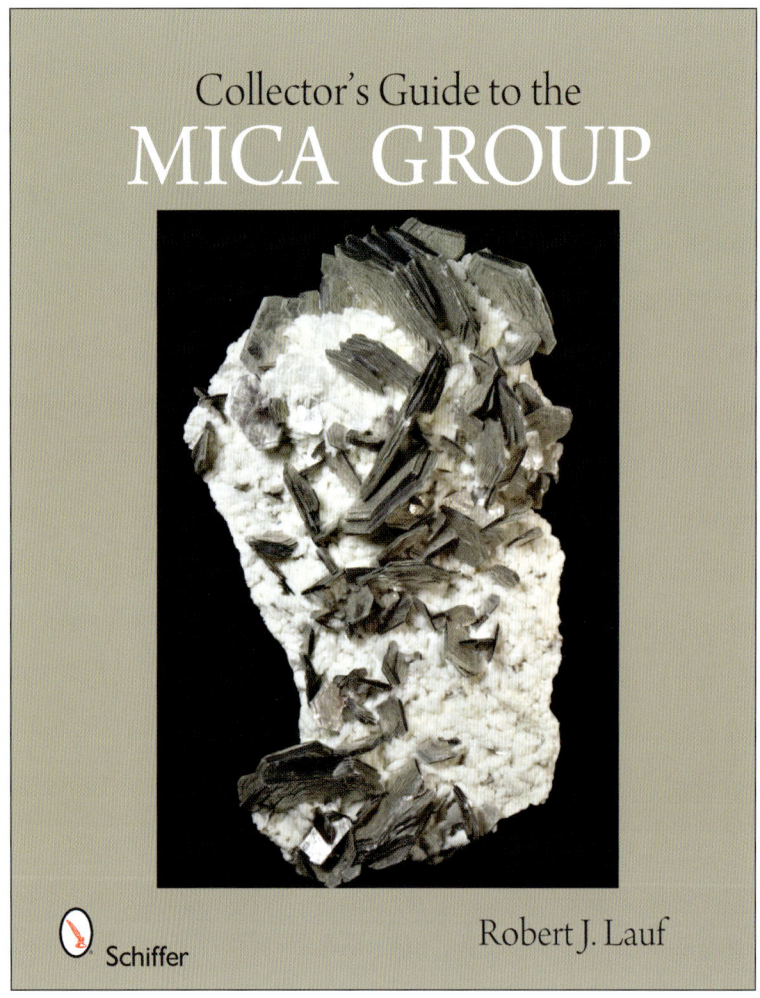

Collector's Guide to the Mica Group. Robert J. Lauf. Mica is a broad term encompassing about forty minerals, ranging from the common to the rare, many at times forming excellent crystals jewelers use. This book features examples recently described among the 115 striking color photos and electron micrographs that illustrate the text. A detailed entry for each type includes information on where each is found, associations of micas with other minerals, pseudomorphs (minerals that masquerade as mica), and micas that fluoresce under UV light. This fascinating guide is for those interested in minerals.

Size: 8 1/2" x 11"   115 color photos   96pp.
ISBN: 978-0-7643-3047-6   soft cover $19.99